D1161730

SCIENTIFIC METHOD

SCIENTIFIC METHOD

THE HYPOTHETICO-EXPERIMENTAL LABORATORY PROCEDURE OF THE PHYSICAL SCIENCES

by

JAMES K. FEIBLEMAN

Tulane University

MARTINUS NIJHOFF / THE HAGUE / 1972

ISBN 90 247 1200 9

PRINTED IN THE NETHERLANDS

PREFACE

In this book I have tried to describe the scientific method, understood as the hypothetico-experimental technique of investigation which has been practiced so successfully in the physical sciences.

It is the first volume of a three-volume work on the philosophy of science, each of which, however, is complete and independent. A second volume will contain an account of the domain in which the method operates and a history of empiricism. A third volume will be devoted to the philosophy of science proper: the metaphysics and epistemology presupposed by the method, its logical structure, and the ethical implications of its results.

Any attempt to describe the experimental method of discovery, as practiced in the physical, chemical and biological sciences, must encounter many difficulties as well as owe its details to many sources. A comprehensive effort is the work of an age; in so far as this book fulfills its ambitions, it represents the labors of many men, and the author's contribution appears chiefly in the point of view from which the abundance of material was organized. A decision to dispense with footnotes was reached only after considering how widespread was the debt owed to others. There were conversations with scientists, and innumerable sources were consulted. An obligation must be acknowledged to Alexis Carrel and to B. Efron as well as to many others whose opinion proved so stimulating and suggestive, notably J. L. Edwards, Albert Einstein, R. I. Dorfman, W. L. Duren, Jr., F. Gonseth, B. J. Pettis, and A. J. Reck. They are not responsible for the final opinions expressed here.

No doubt as the result of much reading it is possible that sometimes in the following pages ideas have been borrowed and assimilated unwittingly, and then presented in a light that might allow them to be thought novel. If so, apologies are in order. On the other hand, a claim to originality has to be put forward for the scheme of the whole as well as for many of the details. The systematic and the intuitive often rub shoulders indiscriminately here.

There remains only the obligation to thank those who have helped me with specific suggestions and the editors who have kindly granted permission to reprint material which first appeared in the pages of their journals. To the former group belong Alan B. Brinkley and Max O. Hocutt. Portion of chapters I and VI were published in *Philosophy of Science;* of chapters IV and V in *Perspectives in Biology and Medicine;* of chapter VIII in *Dialectica;* of chapter IX in *The British Journal for the Philosophy of Science*; and of chapter XIII in *Synthese.*

J.K.F.

New Orleans, 1971

TABLE OF CONTENTS

CHAPTER I

INTRODUCTION: METHOD, DOMAIN AND FINDINGS

1. *The understanding of science*

In this book I will try to describe the method by which the experimental scientists explore the universe of matter and energy. A description of what it is that the physicists and chemists actually do when they are engaged in the enterprise of discovery has to serve as the model for the attempt to formalize that procedure. For when we say "the scientific method" we mean that method which is followed professionally by the scientist in action. For science is nothing static but instead a sequence of dynamic activities intended to investigate some segment of nature. The scientist makes observations, designs instruments, conducts experiments, collects data, takes measurements, makes calculations, draws conclusions. But if we would like to include in the account the process whereby he discovers hypotheses we are hardly ready with an accurate and precise description, for such inductions are difficult to explain. This book purports to be an essay in that direction, nothing more. If it does not help the beginner in science to know where he is headed, it might at least tell the humanist what the scientist is up to.

In order to understand science, therefore, it will be necessary to observe how the scientist works when he is engaged in making discoveries. The description given by the scientist himself is important, but we should not be limited to it, for it often happens that he is better at operating the scientific method than he is at describing it. We need in any case to watch how he works, and then we need to make a qualified interpretation of what we have seen. The method presumably consists in a peculiar technique of investigation, but it does not stand alone. The segments of the natural world it explores and the conclusions it reaches also have to be taken into the account.

These are theoretical considerations. It is not an easy thing to say what the scientists are doing or what science itself is. Experimental science does not as a rule deal in theories which refer to its own method of procedure.

Analysis occurs usually as a stage on the way toward synthesis. Not only is it true that the investigation of details must precede any large-scale conception of the world as disclosed to experience, but it is true also that the successful practice of the scientific method was necessary before any abstraction of it could be undertaken.

It has been claimed that the greatest discovery in science was the method of discovery, yet there is no general agreement as to just what this method is. Every description of such a method is part of a speculative inquiry over which there is much difference of opinion. On what grounds is it to be decided whose description is the correct one? According to some versions, there are many scientific methods; according to others (this one included) there is a single method which has to be modified to suit the extenuating circumstances in almost every instance of its application.

Science itself has become a vast institutional undertaking, receiving the support of many who trust it on faith without comprehending it through reason. Although its successes are well known, the understanding of its precise nature remains vague. Yet faith in the efficacy of science ought not to preclude understanding. Perhaps no one has a grasp of the entire truth but instead each inquirer makes his own contribution. The philosophy of science ought to have room for diverse interpretations as (if nothing more) a device for insuring freedom of inquiry and so for holding open the possibility of progress.

In this book, then, is to be found one more account of science and its operation. The version proposed here was adopted somewhat in the following manner. A study of the current practices in the experimental method of investigation was undertaken in order to discover if possible what is permanent about it, what has survived and is capable of continuing to survive from application to application. The material was developed by observing actual procedures in science – what the scientists do – and by trying to abstract from such observations the formal structure underlying the procedures. It was empathically *not* developed by endeavoring to impose on those procedures a preconceived pattern. Inevitably, the organization of experience rests on patterns gradually acquired from previous experiences; they are not to be confused with those intuitions which proceed without benefit of experience. In the following pages the activity of the scientist is held to be basic, on the assumption that there is no arguing with fact. There is, however, the task of description as well as of interpretation, and the leftover question of how to decide which version of the scientific method is the most acceptable.

The scientific method for making discoveries is a method of human action at the concrete level of experience. It is stated in terms of whole performances of a qualitative as well as a quantitative character, and can therefore not be reduced altogether to symbols. The quality of actually doing science – what it feels like to be carrying out an experiment in the midst of some stage of scientific development – cannot be communicated any more than any other quality can. Nothing substitutes adequately for an act of experience. All that can be done is to suggest what the structure underlying such an experience must be.

It is to be doubted whether any professional scientist ever began his career by attempting to comprehend abstractly the nature of research. The working scientist has for the most part learned his techniques by serving as an apprentice. If for instance we were to ask the technician in a laboratory whether he thought he was engaged in the business of extracting laws from the behavior of matter, we would find the question brushed aside in favor of the immediate operation. He is content to rely upon the intuitively-accepted meaning of science as the possession not so much of a certain kind of knowledge as of the method of obtaining it. We, however, face another problem. We wish to understand this method, so our concern will be with the logical structure underlying the sequence of his activities. This means that we shall have to be at once more explicit, even at the beginning, and more abstract.

No text, of course, can substitute for the actual experience of a working scientist. All descriptions are inherently general and all activities inherently individual, so that description at once says too much and too little. But on the other hand, no scientist is occupied with more than one science; only a specialist in scientific generalities can relate the various sciences both with respect to their method, their field of endeavor and their findings. The attempt to describe the laboratory experience in the more successful physical sciences has to meet many difficulties. Can the logical structure of the scientific method (assuming that there is such a thing) be set forth abstractly when it is always practiced concretely? Science is above all an activity, and has an operational character. It was designed to induce the material world to disclose itself under the form of a group of interconnected abstract structures, to facilitate the discovery of the invariants of that world.

What is stable about science is its method of inquiry. There are many refinements in technique and equipment but the logical structure of the procedure, the multi-stage process of investigation, remains. The field in which those investigations are conducted is, in a word, nature, and nature, too, is here to stay; it changes, but there are many permanent elements as well. The findings of science, however, are tentative, temporary, and subject to alteration, re-evaluation or abandonment. Yet the method exists only for the sake of the findings. This is the paradox of science, that it seeks for what is permanent amid change but is able to make changes in its conclusions as to what is permanent.

The position advanced here actually rests upon an intuition concerning the structure of research and of the axioms from which such a structure could be said to follow logically. The argument is eclectic, in the sense that it owes something to many other inquiries, but attempts a synthesis on the basis of an original position in the metaphysics of methodology. From this position it is understood that the order of substances (as in the case of empiricism) and the order of logic (as in the case of mathematics) are equal in reality though separate, and that attempts to learn about matter through a technique of investigation which involves mathematics as much as it does instrumental extensions of the ordinary senses enables the scientists to elicit reasons.

2. *The definition of science*

The term, science, has been used and misused in every possible connection. Despite the widespread practice and prestige of science, there is no official definition and no one seems to know exactly what science is. It has been understood intuitively and practiced successfully in the physical sciences. There, it is the name for the experimental search for the laws of nature. As understood and practiced, it is the method of field observation and laboratory experimentation, of instrumental trial and mathematical formulation, whereby hypotheses are discovered and tested until they can be established sufficiently to serve for purposes of prediction and control.

The scientific method itself is a sequence of activities, of inductions from concepts derived from facts and supported by experiments to test hypotheses before fitting them into deductive systems. If we define science as the scientific method, and the scientific method as the method of eliciting reasons from substances, we are still left with the necessity of defining substance. Substance may be defined as the irrational ground of individual reaction; it is the genus whose species are matter and energy. Matter and energy are of course interconvertible, matter is static energy, energy dynamic matter. It happens that among the most impressive disclosures for which the scientific method has been responsible is that matter is not the simple stuff the philosophers from Democritus to Feuerbach thought but a very complex and highly powerful set of structures. "Eliciting reason" from an "irrational ground" requires interpretation and makes difficulties for applied mathematics. Scientific method provides a set of basic rules of procedure for arriving

Matter = structure

at decisions concerning the data resulting from observation and experiment.

The scientific investigator is called upon to exercise all of his faculties; he feels, thinks and acts. He feels in terms of sensory observations, thinks in terms of mathematical calculations, and acts in terms of experiments performed. Moreover, he does all these things together, for even when he does them separately each is undertaken in the perspective of its connections with the others. Scientific method insures that the operations of the human faculties shall be externalized in terms of activities and results. The method, then, includes applied logic but contains more. Performing experiments means taking action, and applied logic does not as such necessarily include action. Induction describes the logic of the imagination, but not the act of imagination; otherwise it would be possible to teach investigators how to have original ideas, since they can be taught how induction operates.

3. *The principal divisions of science*

Science for most people means either the romance of the laboratory or the more dramatic applications, either a scientist in a white coat bending over a test tube or a television set and the wonder drugs.

The attention of the professional scientist, however, is spread over three areas which are closely related. These are: the method, the field of investigation, and the findings; or, in other words, science, nature, and laws. There are of course still further considerations, such as for instance the scientific philosophy or the origins of empiricism, but these and like matters will be treated in further volumes. When the word, science, is used, it is not always clear just which of these three areas is meant: the procedure of investigation, the natural world in which the investigation takes place, or the results of such an investigation. But where it is to be shown how things are properly related they must first be clearly distinguished. It often happens that serious misunderstandings and crippling limitations are occasioned by the confusion between two of these three areas. Therefore it may be helpful to consider the distinctions more carefully.

The scientific method is the method of the investigation of nature by means of observation, induction, hypothesis, experiment, calculation, prediction and control. It consists in a set of trained men, called scientists, together with a set of laboratories and equipment, engaged in applying the experimental method in some one of its various stages.

The scientific field is nature, or the world of matter and energy with its multi-layered integrative levels, containing whatever is disclosed to experience by means of sensation, reason and action, whether it came about as the result of human invention or not: stones and trees, animals and planets, space, time, and the galaxies, but also cities and human behavior patterns.

The scientific findings are the laws and other objects which have been found in nature as the result of applying the scientific method. The laws may be causal or statistical, and the objects may be anything from the smallest entity or process to the largest area of empirical investigation, such as nuclear forces, material entities and processes, formulas and rules.

Each science selects some one area of nature for its investigations, and each scientist selects some one sub-area. There are few projects, as there are few laws, which embrace the whole of a scientific domain. Few physical laws embrace the entire physical world, although entropy does, and only one biological law purports to cover all of biology: evolution. Most investigators are content to explore some small segment: the frequency of cosmic rays, the character of the chemical bond, the constituents of the blood, the behavior of man under stress.

The rather precise task of applying the scientific method in special areas has led some specialists to suspect that there is no such thing as the scientific method but that instead there are many methods. There is some truth to the contention. Techniques which have proved so useful in biology will not work in astronomy. It is possible to divide a group of hamsters which are alike in all ascertainable respects, and then to treat both sets alike in all respects save one and then to observe the variations as the results of that one. It is not possible, however, to bring samples of astronomical objects: stars, planets, galaxies, into the laboratory for detailed examination in the same way. Thus there are differences which are forced upon the application of the scientific method and its practitioners by differences in the character of the field investigated – clear differences of the levels in the natural world.

But this is not to say that there are differences in the scientific method itself, or that the differences of levels in the natural world compel the application of different methods. Nature is all of a piece, and the differences in levels, the differences, say, between the physical and the chemical levels or between the chemical and the biological, are not to be taken as important divisions in nature. They are rather fractures or striations where the natural world is found to divide when it is examined; nature is one unbroken structure and the various strata serve to connect as well as to divide.

The differences in treatment are not necessitated by nature but rather by human limitations in the practice of the scientific method. That a cell colony can be brought into the laboratory where a star cluster cannot is an experimental limitation imposed upon the investigator by a difference inherent in the comparative sizes of their objects and not a kind of distinguishing characteristic of the two natural objects which might make one more amenable to the scientific method than the other in such a way that the kind of laws could be found in one that could not be found in the other. The point can be settled, perhaps, by compromise. There is only one scientific method but that method has to be applied with appropriate modifications contingent upon the problems peculiar to the investigation of a particular subject-matter. What justifies writing a book about the scientific method as though there were one and only one method is that the similarities between different applications in different sciences are greater than the differences.

The life of science is preserved over the centuries by just those similarities which are the reference of the term. Nature itself changes continually, as do the scientists, their equipment and techniques, to say nothing of the findings of science, which are subject to continual revision. What remains the same can be sorted out: in nature the invariants in the rate of change, in science the method, in the findings that there are laws. Science is safe as long as it clings to its method, maintains its detached and pure interest in nature, and remains unafraid to abandon or revise its findings. But this cannot be done unless a sharp separation between the three principal divisions is sedulously maintained.

4. *The multi-stage process*

Full abstraction discloses that there is only one scientific method. Although its employment in the separate experimental sciences is always mediated by extenuating circumstances, essentially the same set of procedures, conducted in approximately the same order, can be discovered in laboratory practices. The scientific method is an on-going process, which nevertheless lends itself to analysis into seven well-defined stages. These stages are: observation, induction, hypothesis, experiment, calculation, prediction and control. Each of these stages, except the first, emerges logically from the one before, and each, except the last, leads logically into the next.

Observations are made in order to uncover provocative facts. This stage is one of purely descriptive knowledge. Inductions from the provocative facts are made next in order to discover hypotheses worthy of investigation. This is the stage of brilliant originative insights. The hypotheses are set up for testing in this fashion, then they are tested in three ways. The first way is the one peculiar to science; it involves testing the hypotheses by means of experiments. In the second way the hypotheses are tested against existing theories by means of mathematical calculations. This is the stage in which the quantitative laws are shown to be the necessary logical consequences of a few axioms or assumptions. Finally, the third way is to make predictions from the hypotheses and to use them as instruments to exercise control over practice. Those hypotheses which pass all three tests successfully are considered to be established, however tentatively, as laws.

Not every working scientist engages in all seven of these activities, but the sequence of activities nevertheless is maintained. The multi-stage process is a description of the overall method. It can be and usually is repeated over and over, for it is a self-corrective method. But a word of caution is necessary here. There is nothing automatic about the procedure. The formalized multi-stage process of the scientific method of investigation is a logical structure. But like all logical structures it needs to engage the intuition and constitutes an aid to it, not a substitute for it. We do not have the method complete unless we understand the role of intuition at every stage.

In the following pages a chapter will be devoted to each of these stages.

5. *Beyond the mesocosm*

We shall need to take cognizance of the fact that science employs instruments and techniques to penetrate beyond the range of common experience. There is a little doubt that this constitutes an important contribution. Before the advent of experimental science men were confined in their investigations to that part of the natural world which lay within the reach of the unaided senses. We shall name this available environment the "mesocosm." It employs ordinary reasoning, as guided by syllogistic logic, and extends only to relatively close physical contact with gross objects: the touch of stones and trees, the sounds of dogs and thunder, the sight of the visible stars. The early investigators of nature, such as the Greeks of the fifth century B.C., went as deeply as their unaided senses could take them into the analysis of the contents of the mesocosm. But in the seventeenth century of our era a way was found of starting with ordinary things and arriving at the acquaintance and description of extraordinary things. At first, scientific endeavor adhered fairly closely to the phenomena of middle size which was available to gross experience, the common experience as abstracted and ordered by Newtonian mechanics. But from there it was only a short distance to a knowledge of laws and relationships, of entities and processes, inaccessible to common experience.

Science since then has been occupied with the exploration of two areas adjacent to the mesocosm; the world of the very small: the microcosm, and the world of the very large: the macrocosm. These adjacent worlds are vast ones. The microcosm extends all the way from the quantum of energy and possibly below to the structure of the organic cell, a large spectrum indeed. The macrocosm is equally tremendous, and extends from single stars to the metagalaxy. Good examples of recent scientific developments in these areas would be quantum mechanics for the microcosm and radio astronomy for the macrocosm.

Studies in both the microcosm and the macrocosm began with experience gained in the mesocosm. On the evidence of common experience Priestley and Lavoisier would never have learned about the existence of oxygen nor would Cavendish have discovered that water is a composition of two gasses. The scale of common experience gives the dimensions where the scientific method begins but not where it functions best, for it moves quickly to another analytic level. Common experience takes place on a scale roughly intermediate between one commensurate with the interior of the atom and another commensurate with the exterior of the galaxy. But mere size it not the only indication of the differences between worlds. The mesocosm, for instance, possesses texture in a way the other two do not. For beyond stones and trees it is now possible to find qualities in such objects as electrons at one end of the spectrum, and vast hydrogen clouds at the other.

How the scientific advance was accomplished is often debated, but it would seem to have taken place somewhat as follows.

The European philosophers failed in their efforts to bring together reason and sense experience subjectively; the British empiricists endeavored to account for reason alone, and the Continental rationalists endeavored in the same way to account for sense experience. They failed but in failing pointed up the problem. It appears now that their failure was due to the fact that neither had accounted for action. As we have already seen, the human faculties are not two but three: thought, feeling and action; whereas they had endeavored to account for only one of two: thought or perception.

In contrast with the philosophers, the early European scientists contrived in practice a method of action which incorporated thought and sense perception, the scientific method. The scientific method combines outside the human body the thoughts and sense perceptions which the philosophers especially Kant, failed to get together inside the body. It had not proved possible to incorporate action subjectively; by its nature action is external, overt and objective. Hence the synthesis of the human faculties had to be achieved by extending them into the external world: thought was extended through the formalization of mathematics, feeling was extended by the employment of instruments. When these were engaged together in action by means of experiment, the result was a unified method which for the first time called upon all of the faculties of the human being.

Thus the effort to expend total energies through combined capacities led to the expansion of the available environment and the penetration of the worlds adjacent to the mesocosm, the world opened up by electron microscopes and cyclotrons or by optical and radio telescopes. This had had one important result, among others. For although science takes its departure from the observations conducted at the level of common experience and ends with achievements whose effects are felt at this same level, it goes rather deeply into the adjacent levels. Science characteristically operates beyond the mesocosm.

More recently another development has taken place. There has been an active personal and physical penetration beyond the mesocosm at least in one direction: into the macrocosm. There have been successful attempts to go beyond the narrow confines of the earth's surface, into the ocean by means of the aqualung, and into the space beyond the earth by means of the rocket. From these unique locations still further applications of the scientific methods are possible. They result from the use of the method but they open up the possibility of accelerating its progress.

6. *The interpretation of science*

Science, then, for the most part is named after its method. The novel element in the scientific method is of course experiment. But the hypothetical element is so crucial to the method as set forth here that it could very well be called the hypothetico-experimental method.

The scientific method is a multi-stage process which lends itself to analysis by means of a conceptual framework. The result of this analysis purports to be a new and versatile approach to the scientists' problems and procedures. But it should be remembered that the understanding and interpretation of science itself constitutes a speculative field. While scientists may agree among themselves about the operation of science, they do not agree about its presuppositions, the meaning of its method, or the interpretation of its findings. Differences of opinion in these areas are still to be tolerated.

The present work constitutes a single systematic hypothesis in that field. From this hypothesis, as usual deductions can be made which should be in agreement with the behavior of scientists as well as with the results of science, and of course with the observed facts. But the work is at the same time elementary; it roughs out in pictureable form the main outlines of the scientific procedure.

Only an introductory treatise can hope to touch on all of the stages in the sequence of scientific investigation. An advanced treatise can discuss intensely topics concerned with particular stages along the way; for instance, chance versus determinism, the use of models, the nature of induction or of causation. These are of course important, and they are the topics for which an introductory work must prepare the ground. But at the same time, it is only in an introductory treatise that the whole sweep of the method can be comprehended. Most of the problems discussed here then, are not probed deeply but passed over in the interest of comprehending the method as a whole. It is hoped that other volumes can be provided subsequently which will discuss in more detail some of the questionable points.

A number of interpretations of science are current. Two, however, are outstanding: one which is being deserted and another which is rapidly coming into fashion. It might help the reacher to remind him of what these two are and how the present interpretation is related to them, for it professes to learn from both but to stand on its own ground as a separate third position.

The two interpretations of science correspond closely to two periods in the history of science. The former ran from Galileo through Newton, while the latter has been prompted by the developments of modern physics, with Planck and Einstein. The former, which is the classical interpretation, featured absolutes, causality and determinations. The latter, the modern interpretation, features relative frequencies, probability and indeterminacy. The former was tinged with theological affirmations, the latter is colored by subjectivisms: either the laws of science were inflexible or they depend for their being upon the investigator; either the immutable truth was well known or there are no such truths to be known; either scientific progress made irrefrangible gains or it can never say anything with certainty.

The account of science set forth here means to disclose an interpretation differing from the absolutism of the classical and from the subjectivism of the modern. At the same time, it will be necessary to include what is good about both. Each is to some extent true in what it affirms and false in what it denies. Briefly, the contention is this, that the absolute deterministic causality of the classical interpretation has to be posited but remains in the background, while the statistical indeterministic probability of the modern interpretation approaches it as a limit. Analogically speaking, the classical interpretation posits the role of parameters behind the modern interpretation. But this gives remarkably different results, and changes noticeably the readings of the details. Rationalism and empiricism properly conceived are not in conflict but are correlatives in the interpretation of science, much in the same way that mathematics and instruments work together in actual scientific operations. Just what all this amounts to will be made plain as the exposition proceeds. The understanding of nature as the field in which the scientific method operates, and the findings of the method in that field, are equally influenced by the kind of interpretation which is adopted for the method.

7. *The theory of practice*

It is not the business of scientists to investigate just what the business of science is. Yet the business of science is in need of investigation, and for this purpose we shall need working definitions. Once stated, these definitions may seem an elaboration of the obvious and an over-simplification. But the elaboration often seems obvious only after it has been stated, and the definitions may have to be simple in order to bring out the necessary distinctions.

By "pure science" or "basic research" is meant the investigation of nature by the experimental method in an attempt to satisfy the need to know. Many activities in pure science are not experimental, as, for instance, biological taxonomy, but it can always be shown that in such cases the activities are ancillary to experiment; in the case of biological taxonomy for instance the classifications are of experimental material. Taxonomy is practiced in other areas where it is not scientific, such as in the operation of libraries.

By "applied science" is meant the use of pure science for some practical human purpose.

Thus science serves two human purposes: to know and to do. The former is a matter of understanding, the latter a matter of action. In science, looking for practical results does not come first. The knowledge of the laws has priority: there must be something to apply. Technology which began as the attempt to satisfy a practical need without the use of pure science, will receive a fuller treatment in a later section of this chapter.

Applied science, then, is simply pure science applied. But scientific method has more than one end; it leads to explanation and application. It achieves explanation in the discovery of laws, and then and only then can the laws be applied. Thus both pure science and applied science have aims and results. Pure science has as its aim the understanding of nature, it seeks explanation. Applied science has as its aim the control of nature, it has the task of employing the findings of pure science to get practical tasks done. Pure science has as a result the furnishing of laws for application in applied science. And, as we shall learn later in this chapter, applied science has as a result the stimulation of discovery in pure science.

Applied science puts to practical human uses the discoveries made in pure science. Whether there would be such a thing as pure science alone is hard to say; there are reasons for thinking that there would be, for pure science has a long history and, we have noted, another justification. There could be technology without science, for millennia, in fact, there was. But surely there could be no applied science without pure science, applied science means just what it says, namely, the application of science, and so without pure science there would be nothing to apply.

Logically, pure science pursued in disregard of applied science seems to be the *sine qua non* of applied science, while historically the problems toward which applied science is directed came before pure science.

It has been asserted, for instance, that Greek geometry, which is certainly pure, arose out of the interest in land surveying problems in Egypt, where the annual overflow of the Nile obliterated all conventional boundaries. Certainly it is true that the same concept of infinity is necessary for the understanding of Euclidean geometry and for the division of farms. Be that as it may, it yet remains true that the relations between pure and applied science are often varied and subtle, and will require exploration.

Let us propose the hypothesis that all pure science is applicable.

No proof exists for such an hypothesis, all that can be offered is evidence in favor of it. This evidence consists of two parts, the first logical and the second historical.

The logical evidence in favor of the hypothesis is contained in the very nature of pure science itself. Any discovery in pure science that gets itself established will have gained the support of experimental data. Thus there must be a connection between the world of fact, the actual world, in other words, which corresponds to sense experience, and the laws of pure science. It is not too difficult to take the next step, and so to suppose that the laws, which were suggested by facts in the world corresponding to sense experience, could be applied back to that world.

The second part of the evidence for the applicability of all of the laws of pure science is contained in the record of those laws which have been applied. Almost everything in the modern western nations has been altered by applied science, and now Soviet Russia and Communist China are following their lead.

Indeed, so prevalent are the effects of applied science, and so concealed the leadership of pure science, that those whose understanding of science is limited are apt to identify all of science with applied science and even to assume that science itself means technology or heavy industry.

One argument against the position advocated here would be based upon the number of pure theories in science for which no application has yet been found. But this is no argument at all, for to have any weight, it would have to show not only that there had been no application but also why there could be none. Yet a time lag between the discovery of a theory and its application to practice is not uncommon. How many years elapsed between Faraday's discovery of the dynamo and its general manufacture and use in industry? Conic sections were discovered by Appollonius of Perga in the third century B.C., when they were of intellectual interest only, and they were applied to the problems of engineering only in the seventeenth century of our era. Non-Euclidean geometry, worked out by Riemann as an essay in pure mathematics early in the nineteenth century, was used by Einstein in his theory of relativity in the twentieth century. The coordinate geometry of Descartes made possible the study of curves by means of quadratic equations. It was in Descartes' time that the application of conic sections to the orbit of planets was first noticed; later the same curves were used in the analysis of the paths of projectiles, in searchlight reflectors, and in the cables of suspension bridges. Chlorinated diphenylethane was synthesized in 1874. Its value as insecticide (DDT) was not recognized until 1939 when a systematic search for moth repellants was undertaken for the military. The photoelectric cell was used in pure science, notably by George E. Hale on observations of the sun's corona in 1894. Twenty-five years later it was found employed in making motion pictures. It often happens that the discovery of a useful material is not sufficient; it is necessary also to discover a use, to connect the material to some function in which it could prove advantageous. Paracelsus discovered ether and even observed its anaesthetic properties, and Valerius Cordus gave the formula for its preparation as early as the sixteenth century. But it was many centuries before ether was used as an anesthetic.

Presumably, then, pure scientific formulations which have not been applied are merely those for which as yet no applications have been found. In the effort to extend knowledge it is not strategically wise to hamper investigation with antecedent assurances of utility. Many of the scientific discoveries which later proved most advantageous in industry had not been self-evidently applicable. This is certainly true of Gibbs' phase rule in chemistry, for instance. It often happens that for the most abstract theories new acts of discovery are necessary in order to put them to practical use.

8. *From theory to practice*

It should be observed at the outset that applications are matters of relevance. The line between pure and applied science is a thin one; they are distinct in their differences, but one fades into the other. For instance, the use of crystallography in the packing industry is an application, but so is the use of the mathematical theory of groups in pure crystallography and in quantum mechanics. The employment of mathematics in pure science means application from the point of view of mathematics but remoteness from application from the point of view of experiment. Some branches of mathematics have been so widely employed that we have come to think of them as practical affairs. This is the case with probability or differential equations. Both branches, however, considered in their mathematical aspects and not at all in relation to the various experimental sciences in which they have proved so useful, are theoretical disciplines with a status of their own which in no wise depends upon the uses to which they may be put.

Procedurally, the practitioner introduces into his problem the facilitation afforded by some abstract but relevant theory from either mathematics or pure science. The statistical theory of extreme values is a branch of mathematics, yet it has application to studies of metal fatigue and to such meteorological phenomena as annual droughts, atmospheric pressure and temperature, snowfalls, and rainfalls. The discovery of the Salk vaccine against polio virus was an achievement of applied science. Yet Pasteur's principle of pure science, that dead or attenuated organisms could induce the production of antibodies within blood serum, was assumed by Salk, so the immensely important practical application would not have been possible without the previous theoretical work.

It is clear, then, that we need three separate and distinct kinds of pursuits, and, perforce, three types of interest to accompany them. The first is pure science. Pure theoretical sciences are concerned with the discovery of natural law and the description of nature, and with nothing else. These sciences are conducted by men whose chief desire is to know, and this requires a detached inquiry – which Einstein has somewhere called "the holy curiosity of inquiry" and which Emerson declared to be perpetual. Such a detachment and such a pursuit are comparable in their high seriousness of purpose only to religion and art.

The second type is applied science, in which are included all applications of the experimental pure sciences. These are concerned with the improvement of human means and ends and with nothing else. They are conducted by men whose chief desires are practical: either the improvement of human conditions or profit, or both. Temperamentally, the applied scientists are not the same as the pure scientists: their sights while valid are lower; they are apt to be men of greater skill but of lesser imagination; what they lack in loftiness they gain in humanity. It would be a poor view which in all respects held either variety secondary. Yet there is a scale of order to human enterprises, even when we are sure we could dispense with none of them; and so we turn with some measure of dependence to the type of leadership which a preoccupation with detached inquiry is able to provide.

The third type is the intermediate or *modus operandi* level, which is represented by the scientist with an interest in the solution of the problems presented by the task of getting from theory to practice. As Whitehead said, a short but concentrated interval for the development of imaginative design lies between them. Consider for example the role of the discovery of Hertzian waves, which not only led to the development of radio but also brilliantly confirmed Maxwell's model of an electromagnetic field, especially the existence of electromagnetic waves, with a constant representing the velocity of light in two of the four equations.

The conception of science as exclusively pure or utterly applied is erroneous, the situation is no longer so absolute. When scientific theories were not too abstract, it was possible for practical-minded men to address themselves both to a knowledge of the theory and to the business of applying it in practice. The nineteenth century saw the rise of the "inventor," the technologist who employed the results of the theoretical scientist in the discovery of devices, or instruments, of new techniques in electro-magnetics, in chemistry, and in many other fields. Earlier scientists like Maxwell had prepared the way for inventors like Edison. In some sciences, notably physics, however, this simple situation no longer prevails. The theories discovered there are of such a degree of mathematical abstraction that an intermediate type of interest and activity is now required. The theories which are discovered in the physicists' laboratories and published as journal articles take some time to make their way into engineering handbooks and contract practices. Some intermediate theory is necessary for getting from theory to practice.

A good example of the *modus operandi* level is furnished by the activities and the scientists concerned with making the first atomic bombs. Hahn and Strassmann discovered in 1938 that neutrons could split the nuclei of uranium. Einstein and Planck had earlier produced the requisite theories, but it was Enrico Fermi, Lise Meitner, and others who worked out the method of getting from relativity and quantum mechanics to bombs which could be made to explode by atomic fission.

9. *Technology*

There has been some misunderstanding of the distinction between applied science and technology; and understandably so, for the terms have not been clearly distinguished. Primarily the difference is one of type of approach. The applied scientist as such is concerned with the task of discovering applications for pure theory. The technologist has a problem which lies a little nearer to practice. Both applied scientist and technologist employ experiment; but in the former case guided by hypotheses deduced from theory, while in the latter case employing trial-and-error or skilled approaches derived from concrete experience. The theoretical biochemist is a pure scientist, working for the most part with the carbon compounds. The biochemist is an applied scientist when he explores the physiological effects of some new drug, perhaps trying it out to begin with on laboratory animals, then perhaps on himself or on volunteers from his laboratory or from the charity ward of some hospital. The doctor or practicing physician is a technologist when he prescribes it for some of his patients.

Speaking historically, the achievements of technology are those which develop without benefit of science; they arose empirically either by accident or as a matter of common experience. The use of certain biochemicals in the practice of medicine antedates the development of science: notably, ephedrine, cocaine, curare, and quinine. This is true also of the pre-scientific forms of certain industrial processes, such as cheese-making, fermentation, and tanning.

The applied scientist fits a case under a class; the technologist takes it from there and works it out, so to speak, *in situ*. Applied science consists in a system of concrete interpretations of scientific propositions directed to some end useful for human life. Technology might now be described as a further step in applied science by means of the improvement of instruments. In this last sense, technology has always been with us; it was vastly accelerated in efficiency by having been brought under applied science as a branch.

Technology is more apt to develop empirical laws than theoretical laws, laws which are generalizations from practice rather than laws which are intuited and then applied to practice. Empirical procedures like empirical laws are often the product of technological practice without benefit of theory. Since 1938, when Cerletti and Bini began to use electrically induced convulsions in the treatment of schizophrenia, the technique of electroshock therapy has been widespread in psychiatric practice. Yet there is no agreement as to what precisely occurs or how the improvement is produced; a theory to explain the practice is entirely wanting.

Like applied science, technology has its ideals. Let us consider the technological problem of improving the airplane, for instance. For a number of decades now, the problem to which airplane designers have addressed themselves is how to increase the speed and the pay load of airplanes. This means cutting down on the weight of the empty airplane in proportion to its carrying while increasing its effective speed. If we look back at what has been accomplished in this direction, then extrapolate our findings into the future in order to discern the outlines of the ideal, we shall be surprised to discover that what the designers have been working towards is an airplane that will carry an infinite amount of pay load at an infinite speed while itself weighing nothing at all! This of course is a limit, and like all such limits, is an ideal intended to be increasingly approached without ever being absolutely reached.

Conception of the ideal is evidently of the utmost practicality and cannot be escaped in applied formulations. Yet the existence of such a thing as a technical ideal is fairly recent and is peculiar to western culture. The ideal of a general character envisaged in this connection is that of fitness of purpose and of economy; no material or energy is to be wasted. Roman engineers built bridges designed to support loads far in excess of anything that might be carried over them: their procedures would be regarded as bad engineering today. The modern engineer builds his bridge to carry exactly the load that will be put upon it plus a small margin for safety, but no more; he must not waste structural steel nor use more rivets than necessary, and labor must be held to a minimum. The ideal of technology is efficiency.

Although technologists work in terms of ideals, they are nevertheless more bound down to materials than is the applied scientist, just as the applied scientist in his turn is more bound down to materials than is the pure scientist. Since the technologist is limited by what is available, when he increases the going availability it is usually at the material level. The environment with which a society reacts is the available environment, not the entire or total environment, and the available environment is that part of the environment which is placed within reach of the society by its knowledge and techniques. These are laid out for it and increased by the pure and the applied sciences. Only when these limits are set can the technologist go to work. For example, discovery of the internal combustion engine which required gasoline as fuel turned men's attention to possible sources of oil. The applied scientist found ways of locating oil in the ground, while the petroleum technologist made the actual discoveries. In the hundred years since Edwin Drake drilled the world's first oil well near Titusville, Pennsylvania, in 1859, the technologists have taken this discovery very far. Oil is now a natural resource, a part of the available environment, but it can hardly be said to have been so a hundred years ago, although it was just as much in the ground then as now.

Another kind of technologist is the engineer. Engineering is the most down-to-earth of all scientific work that can justify the name of science at all. In engineering the solutions of the technologists are applied to particular cases. The building of bridges, the medical treatment of patients, the designing of instruments, all improvements in model constructions of already existing tools – these are the work of the engineer. But the theories upon which such works rests, such as studies in the flow or "creep" of metals, the physics of lubrication, the characteristics of surface tensions of liquids – these belong to the applied scientist.

The industrial scientific laboratory is devoted to the range of applied science, from "fundamental research," by which is meant long range work designed to produce or improve practical technology, —to immediate technological gains from which manufacturing returns are expected, for instance, the testing of materials and of manufactured products. Technological laboratories have been established in the most important of the giant industrial companies, such as DuPont, General Electric, Eastman Kodak, and Bell Telephone. The work of such laboratories is cumulative and convergent, applied science in such an institution is directed toward eventual technological improvement, the range of applied science and technology being employed as a series of connecting links to tie up pure science with manufacturing. University and foundation laboratories often serve the same purpose, but with the emphasis shifted toward the theoretical end of the scientific spectrum.

The development of technology has a strong bearing on its situation today and may be traced briefly. In the Middle Ages, there was natural philosophy and craftsmanship. Such science as existed was in the hands of natural philosophers, and such technology as existed was in the hands of craftsmen. There was precious little of either, for the exploration of the natural world was conducted by speculative philosophers, while the practical tasks were carried out by handicrafts employing comparatively simple tools, although there were exceptions: the windmill, for instance. There was little commerce between them, however, for their aims were quite different, and the effort to understand the existence of God took precedence over lesser pursuits.

Gradually, however, natural philosophy was replaced by experimental science, and handicraft by the power tool. The separation continued to be maintained, and for the same reasons; and this situation did not change until the end of the eighteenth century. At that time, the foundations of technology shifted from craft to science. Technology and applied science ran together into the same powerful channels at the same time that the applications of pure science became more abundant. A craft is learned by the apprentice method; a science must be learned from the study of principles as well as from the practices of the laboratory, and while the practice may come from applied science the principles are those of pure science.

There is now only the smallest distinction between applied science, the application of the principles of pure science, and technology. The methods peculiar to technology: trial-and-error, invention aided by intuition, have merged with those of applied science: adopting the findings of pure science to the purpose of obtaining desirable practical consequences. Special training is required, as well as some understanding of applied and even of pure science. In general, industries are based on manufacturing processes which merely reproduce on a large scale effects first learned and practiced in a scientific laboratory. The manufacture of gasoline, penicillin, electricity and oxygen, were never developed from technological procedures, but depended upon work first done by pure scientists. Science played a predominant role in such physical industries as steel, aluminum, and petroleum; in such chemical industries as pharmaceuticals and potash; in such biological industries as medicine and animal husbandry.

A concomitant development, in which the triumph of pure science over technology shows clear, is in the design and manufacture of instruments. The goniometer, for the determination of the refractive index of fluids (used in the chemical industries); the sugar refractometer, for the reading of the percentages by weight of sugar (used in sugar manufacture); the pyrometer, for the measurement of high temperatures (used in the making of electric light bulbs and of gold and silver utensils); the polarimeter; for ascertaining the amount of sugar in urine (used by the medical professsion); these and many others, such as for instance the focometer for studies in the length of objectives, the anomaloscope for color blindness, and the spectroscope for the measurement of wave lengths, are precision instruments embodying principles not available to the technologist working unaided by a knowledge of pure science.

10. *From practice to theory*

In the course of pursuing practical ends abstract principles of science hitherto unsuspected are often discovered. The mathematical theory of probability was developed because some professional gamblers wished to know the odds in games of chance. Electromagnetics stimulated the development of differential equations, and hydrodynamics function theory. Carnot founded the pure science of thermodynamics as a result of the effort to improve the efficiency of steam and other heat engines. Aerodynamics and atomic physics were certainly advanced more swiftly because of the requirements of war. Air pollution, which accompanies big city "smog," has led a number of physical chemists to investigate the properties of extreme dilution. Hence it is not surprising that many advances in pure science have been made in industrial laboratories: from the Bell Telephone laboratories alone have come the discoveries by Davisson and Germer of the diffraction of electrons, by Jansky of radio astronomy, and by Shannon of information theory.

Technology has long been an aid and has furnished an impetus to experimental science. The development of the delicate mechanisms requisite for the carrying out of certain experiments calls on all the professional abilities of the instrument maker. Such a relation is not a new one; it has long existed. The skill of the Venetian glass-blowers made possibly many of Torricelli's experiments on gases. Indeed, glass can be followed through a single chain of development for several centuries, from the early microscopes to interferometers. The study of electromagnetics was responsible for the later commerce in electric power and the vast industries founded on it. But, contrariwise, thermodynamics grew up as a result of the problems arising from the use of steam in industry. We cannot afford to neglect in our considerations the economic support as well as the social justification which industry has furnished, and continues to be prepared to furnish, to research. The extraordinary rise of pure chemistry in Germany was not unrelated to the industry constructed on the basis of the aniline dyes, as well as cosmetics and explosives, during the nineteenth century.

The harm to practice of neglecting the development of pure theoretic science will not be felt until the limits of installing industries by means of applied science and technology, and of spreading its results, have been reached. Science can to some extent continue to progress on its own momentum together with such aids as the accidental or adventitious discoveries of pure science in technological laboratories are able to furnish it. But there are limits to this sort of progress. Thus far the communist countries of the east have taken every advantage of the scientific developments achieved in the capitalist countries of the west. But after all, the applied science which has come out of the west has been the result of its own preoccupations with pure science and with theoretic considerations which lay outside the purview of any practice. Industrial laboratories may occasionally contribute to pure science, but this is not their chief aim, and it is apt to be forgotten that such industries would not exist were it not for the fact that some centuries ago a handful of scientists with no thought of personal gain or even of social benefit tried to satisfy their curiosity about the nature of things. The restless spirit of science, never content with findings, hardly concerned with the applications of findings, is always actively engaged in pursuing methods in terms of assumptions, and must have some corner of the culture in which it can hope to be protected. A wise culture will always offer it elbow room, with the understanding that in the future some amortized inquiry is bound to pay dividends. The ivory tower can be, and sometimes is, the most productive building in the market place.

Of course, applied science and technology cannot be independent of pure science, nor can pure science be independent of applied science and technology. The two developments work together and are interwoven. Gilbert discovered that the freely suspended magnetic needle (i.e., the compass) could be a practical aid to navigation at the same time he proposed that perhaps the earth was a gigantic magnet.

Problems which arise in the midst of practical tasks often suggest lines of theoretical inquiry. But there is more. Pragmatic evidence has always been held by logicians to have little standing. A scientific hypothesis needs more support than can be obtained from the practical fact that "it works." For who knows how long it will work or how well? What works best today may not work best tomorrow. A kind of practice which supports one theory may be supplanted by a more efficient kind of practice which supports quite a different theory. Relativity mechanics gives more accurate measurements than Newtonian mechanics. That use does not determine theory can be easily shown. Despite the theoretical success of the Copernican theory as refined and advanced by Kepler, Galileo, and Newton, we have never ceased to use the Ptolemaic conception in guiding our ships or in regulating our clocks. However, if the practical success achieved by the application of certain theories in pure science cannot be construed as a proof of their truth, neither can it be evidence of the contradictory: workability is no evidence of falsity, either. Newton is still correct within limits. Practicality suggests truth and supports the evidence in its favor even if offering no final proof. The practical uses of atomic energy do not prove that matter is transformable into energy, but they offer powerful support. Hence, the use of a scientific law in the control of nature constitutes the check of prediction and control.

11. *Cross-field applications*

We have treated all too briefly the relations of theory to practice and of practice to theory in a given science. We have also mentioned the productive nature of cross-field research. It remains now to discuss a last dimension of relations between science and practice, and this is what we may call cross-field applications: the employment of the practical effects of one science in those of another.

The applications of science have been greatly aided by the cross-fertilization of techniques. Radio astronomy, which has proved so useful in basic research, being already responsible for the discovery of "radio stars," and for adding to our knowledge of meteor streams and the solar corona, owes its inception to a borrowed instrument. Cross-field application has a long history, dating at least to the early half of the eighteenth century when distilleries in England brought together the results of techniques of producing gin acquired from both chemistry and theories of heat energy. Perhaps the most prominent instance of this is the way in which medicine has drawn upon the physical technologies. The use of the vacuum tube amplifier and the cathode ray oscillograph in determining the electrical potential accompanying events in the nervous system, the entire areas covered by encephalography and by roentgenology show the enormous benefits which have accrued to medical studies and procedures. Scintillation counters, developed and used in physical research, have been employed to measure the rate at which the thyroid gland in a given individual removes iodine from the blood stream; to measure the natural radioactivity of the body; to determine the extent of ingested radioactive compounds in the body; to assay the radioactive iron in blood samples. Chemistry has been an equally potent aid to pharmacology, which would have hardly existed in any important sense without it.

Other instances abound, and indeed multiply every day. In 1948 the Armour Research Foundation sponsored a Crystallographic Center, of interest to pharmaceutical corporations because of the crystalline nature of some of the vitamins. The invention of automatic sequence controlled calculators and other types of computing machines has seen their immediate application to atomic research and to military problems of a technical nature. Medical knowledge is being placed at the service of airplane designers, who must estimate just what strains their aircraft will demand of pilots. Perhaps the most graphic illustration of cross-field application is in scientific agriculture. Here hardly a single science can be omitted: physics, chemistry, biology – all contribute enormously to the joint knowledge which it is necessary to have if soils are to respond to management.

The cross-field applications of science usually work upward in fields corresponding to the integrative levels of the sciences. Applications found in physics will be employed in chemistry or biology, those found in chemistry will be employed in biology or psychology, those found in biology will be employed in psychology or sociology, and so on. The use of physics in biology has in fact brought in to existence the science of biophysics, in which are studied the physics of biological systems, the biological effects of physical agents, and the application of physical methods to biological problems.

The cross-fertilization of applied science, the use of techniques, skills, devices, acquired in one science to achieve gains in another, has effects which tend to go beyond either. They add up to a considerable acceleration in the speed with which the applied sciences affect the culture as a whole. In the brief space of some several hundred years, western culture has been altered out of all recognition by the employment of applied science and technology to the purposes of industry, health, government, and war. Much of the alteration has been accomplished by means of the cooperation between the sciences. We now know that the shortest route to an effective practice lies indirectly through the understanding of nature. If there exists a human purpose of a practical kind, then the quickest as well as the most efficient method of achieving it is to apply the relevant natural laws of science to it.

12. *The aims of this handbook*

There are limits to what a book of this sort can hope to accomplish. Yet from two points of view the treatment it presents seems required. In the first place, the rapid proliferation of work in the sciences, both pure and applied, and the literature accompanying the same, has made it impossible for any one to assimilate all of it by any approach other than an attempt to get at the essential meaning. In the second place, due to the political and military importance of science, there are forces which tend to threaten it or at the very least limit its activities. Some day, mankind having destroyed science may wish to revive it, and then it will be necessary to know what it had been.

Such a study could be important to the future, but this it not all. It is one of the tasks of the philosophy of science to interpret the method, the field and the findings of science in terms of ordinary experience. This has been done often with the findings, but more rarely with the method itself. Science, as we have noted, begins its investigations at the level of common experience but soon develops instrumental and mathematical analyses which reach into levels not available to common experience. What does such penetration mean in terms of common experience, to what extent is it capable of changing common experience, and how is the activity of science to be understood by those who do not practice it and are not trained for it? The discussion in depth of philosophical problems arising in connection with the scientific method is here subordinated to a detailed description of the practice of that method. Now a hazard of what the method of discovery in science is like is a theory, and there is room in any speculative field for more than one theory, at least until the truth can be discovered.

The picture presented in this book, then, is that of a purportedly ideal cycle of scientific investigation. It does not provide for accidental factors which may affect it materially, such as chance discovery or the contributions of applied science and technology to basic research. Such an ideal portrait presents a misleading complete conception of the method, for the stages do not have to be followed in the sequence given or by the same investigator. There are alternative choices to be made at every stage. Science is a cooperative undertaking in which an individual genius may furnish the impetus but not necessarily all of the steps. That is why the examples employed in this presentation have not always been brought up to date but range over the whole history of science. The idea has been not necessarily to acquaint the reader with the latest developments in science but rather to make him somewhat familiar with what sciences as a going concern is like.

At the present time science is taught largely on the apprentice method, and the scientist is not particularly aware that he is following any established procedure; rather is he always occupied with some specific problem involved in that procedure, guided chiefly by past experience, by accepted tradition, training, memory, and imitation. In the laboratories of schools and colleges, in addition to the memorizing of sets of names for entities and processes, and the standard formulations of established laws, the scientific procedures are communicated by means of the endless repetition of certain of the classical experiments. Even thought there is no substitute for concrete experience, taken alone it is crude at best and needs to be supplemented with a manual setting forth the outlines of the method as it is currently employed. This book, it is hoped, will fulfill such a need.

There are many audiences for scientific method; not only those students who are preparing themselves for a productive career in scientific discovery and who need to be reminded at the same time of the independence of theory from practice, but also those humanists who hold an understanding of the experimental sciences to be essential to a complete education. Experimental science ought to be considered along with the humanities; it contains large-scale theories which need to be assimilated. In a proper sense, all disciplines can be listed under the heading of inquiry, and the method which has been so prominent in the physical sciences because of its brilliant successes is also an endeavor to increase the sum of human knowledge.

There are empirical studies in which the scientific discipline has not yet been successfully applied, partly because of the inherent difficulties of a complex subject-matter but also partly because the confusion of the physical sciences with their method has led to an imperfect understanding of the scientific method *in abstracto* – the social studies, for instance. One aim of this book, then, is to abstract the scientific method as employed in the physical sciences in order that it might be available for use in other empirical studies.

What is the role of this book in the physical sciences themselves? Procedures which have nothing more solidly logical to go on than intuitive understanding may easily fail eventually. Yet it is not as though future scientists were to be guided by this treatise; the only hope is that they might allow themselves to be reminded by it.

CHAPTER II

THE SEARCH FOR DATA: OBSERVATION

Scientific method begins with the observation of nature. And not with the whole of nature but only with some part. And what that part is, it is important to examine, and how observation becomes channelled accordingly, we must make efforts to disclose. Considerable care must be taken to ask just the right questions; since from the examination of what it is that scientists do, we know that we shall obtain only a selection of the answers that fall within the limits established by what was chosen as our area of inquiry. We seem the least committed if we start by looking into our working conditions. What, then, is the scientific method?

The description of such a method has to be made in terms of the goals of the scientists. No account of what a man is doing could be adequate without some knowledge of what it is that he is trying to do. Our procedure will be to endeavor to elicit his aims from his practice. We shall see when we are done that we have uncovered the outlines of a logical structure.

We make a start, then, with the world in which he finds himself, and within which he proposes to conduct a certain kind of investigation. We are contemplating a fully-developed method and so we shall be obliged to take into account an advanced science. We shall for this purpose assume the existence within nature of a laboratory well equipped and a scientist well prepared. The laboratory will contain many instruments as well as ready access to the raw materials for constructing others. The scientist will have a thorough familiarity with the existing knowledge in his field, consisting of data covered by laws which have been so far as possible arranged deductively. The equipment of the laboratory and the preparation of the investigator have in common a silent background of principles which themselves rest on covert metaphysical axioms established as beliefs. Examples of such axioms are: that material objects are members of classes, and that the external world is independent of the observer and knowable. The scientist is definite about what in particular he wishes to study but he is not certain of what he may expect to find. Science, under these conditions, begins with observations.

In the practice of the scientific method, the scientist employs all three of his faculties; he observes by means of his senses, but he does not merely observe, he also acts and thinks. By means of the action he contrives to determine to some extent what it is that he shall observe; and by means of the thought he endeavors to interpret what it is that he is observing. We are not here concerned with the psychological aspects of this procedure but with what the procedure itself discloses and with how such disclosures are brought about.

What the scientist observes is of course a material object or event, and he does so in the field or in his laboratory. Here as always he has need of all his faculties but chiefly of his senses. There is a second stage, however, in which action primarily is called for, and this is when he must arrange the object to be observed. A third stage involves the dominance of thought, this is when the observer must interpret the results of his observations. We shall speak first of feeling as (1) *simple observation,* insofar as this is at all possible. Then we shall introduce action to the procedure when we come to consider (2) *controlled observation.* And, finally, we shall emphasize the element of thought in referring to (3) *the facts as observed.*

1. *Simple observation*

By simple observation is meant deliberate exposure to sensory stimuli. Observation is passive but it is not altogether passive, as we have shortly to see, but in its passive phase it is at its simplest. This means taking up a perspective on some segment of the external world. The observer arranges himself in such a way that he will receive particular sense impressions in order to describe them. Thus thought and action play a role in simple observation but a lesser role than feeling.

The kind of single-minded observation required here is neither natural nor accidental; it requires concentration, it requires training. It involves a sort of passion of intensity which does not come easy nor without practice. All of the faculties must be brought to bear to reinforce and support sensory observation, so that what the observer does he does with his whole body, his whole being.

Simple observation includes under the assumptions of pure phenomenology a naive description of what happens. The study of appearances is the effort to become passive in the presence of the recording of sense impressions. But at this point it is necessary to ask why one kind of entity, process or relationship attracts the observer's attention more than another. Observation of a rudimentary sort involves a kind of sensory scanning. The number of material objects existing and available to observation always far exceeds the number observed.

The answer is that the observer occupies a perspective on the object which both permits his observations and conditions them. But casual or arbitrary observations though they exist in science are not the rule in scientic method. Long before there was science there were human beings equipped with sense organs and brains; what, then, made them into scientists? Science is not a matter of mere observation, for the religious practice of divination is also an observational discipline. Seeing a bird at the right moment or in an auspicious location may be "evidence" for a future event; the "good omen" of the diviner is an object observed. It must therefore be something else connected with observation in which the scientific distinction rests, something in the selection of the data to be observed, in the general principles governing the selection of the data, or perhaps in the interpretation itself, that renders science unique.

One of the most important factors in observation is the use of instruments. Even in simple observation it is possible to extend the senses by means of instruments. The senses may be extended, as with the optical telescope, or they may be supplanted, as with the Geiger counter, but in any case the observer observes: he observes the object directly or he observes the pointer readings made on the instrument by means of its exposure to the object. Measurement is a form of observation.

There is no scientific observation without self-discipline; yet thought is employed at this stage to keep observation pure and simple. It is a sophisticated notion that a certain deliberateness is required to preserve the simplicity of observation. If we were to understand by observation a process so simple that only the barest minimum of interpretation was allowed, then, at least in the case of vision, colored shapes would be all that could be actually counted. Gradually, however, as the observer concentrated on the distinction between this colored shape and its immediate environment (which by contrast also limits it), he would perceive that he was dealing with a disjunct object. Next he would perceive that the object perhaps had internal features; and if he were able to associate the discriminated properties with those he was able to remember from similar objects, then he would be ready recognize in his "object" a member of a class.

Thus by means of the similarity of properties of diverse individuals a bridge is made to the generality of classes. The object so-called may be a material thing or process, a tenuous element or a concrete event; but in any instance the scientific observer recognizes the general by means of the individual, and it is the uniformity of generals that he is after. No one would dissect a rat if all he could learn from it were the properties of that particular rat. For no individual is important except as it is typical. The scientist wants to be sure of his ground at all stages, and so he can learn about rats only if he has recourse at some point to simple observation. In science a set of observations is always regarded as a sample of a population, which is indefinitely large and possibly infinite as indicating a class of relations which stand behind the finite observations. The infinite population is and must remain hypothetical; it is posited, yet infinity has never been observed.

There is no such thing as wholly random observation any more than there is naive experience. An observer may think he sees geese flying southward in the fall, a gathering storm, or illness in the face of a friend. Yet these are not simple facts at all, they are highly ingenious inferences. He does not see geese flying southward or even geese flying. Think what is contained in the word "geese" alone from this point of view, a full description of which would require volumes. And the same is true of storms and illnesses, all in a way elaborate constructions which as such are not observed; for here there already is high complication and reliance upon previously acquired interests. From the perspective shaped by earlier experiences, individuals are recognized as members; for instance in all three examples observation has meant fitting cases under classes: comparing *this* storm with storms lived through before, matching *these* geese against previous geese, and fitting *this* illness with other illnesses.

Most important of all perhaps is the question of why he noticed these phenomena rather than others in the first place. The observer's professional attention is caught by inclinations, by signs of entities and processes, by relations, and even by absent objects whose presence was expected. In the field or laboratory, there is always more – phenomenologically speaking – than can be observed, and more, too, than is pertinent to any observational program. Observations always involve a selection, for each of them happens against a background and within a context of other happenings.

One of the chief difficulties of science at the very outset of simple observation is that all experience is particular while all descriptions of experience are general. The language whereby the observer gives an account of his experience is incurably general. All words except the names of singular objects, such as "London" or "Napoleon," are class names, and it is by combining class names into sentences that the observer undertakes to record what he has observed. There is an advantage for science in the fact that in so far as the sentences are general they refer to other and similar situations in the present and future. Thus the descriptions of experience are predictive of the experience of others, and science takes its place as a public undertaking available to continual verification. The observation statements of the observer will, however, stand investigation only to the extent to which they are impartial and impersonal.

The chief requirement in the case of observation is that it shall be as independent as possible of all extraneous considerations. Only those beliefs which are necessary in order to make observation itself possible are allowed. Preconceptions, anticipations, adjacent materials, never completely avoidable, are prime sources of danger to the act of observation if it is to have any scientific value. Observation, in short, must be as naive and as pure as it can be made. It must be largely unaffected by subjective or objective interference, and as much as possible at the mercy of the data. This is what scientists mean by "saving the phenomena." No way of interpreting the observed material so that it means something other than what the observations themselves reveal, is allowed.

Investigations always begin, with certain preconceptions concerning the objects to be investigated, though this is seldom recognized to be the situation by the observer, who always fancies that he is addressing himself unhampered to the task at hand. And the chief of his preconceptions is the supposition (possibly correct, of course) that there are such objects as he proposed to investigate, and, further, that if there are other such objects then that these are the preferred objects.

But there is one peculiarity which is perhaps paramount. The scientist observing is referred by his senses to some datum of experience which has the peculiar property that it is not answerable to the usual place assigned to it in the known scheme of things. The data of observation disclose a number of novelties, and so the observer's role, once he has selected his field of observation, is passive to perception: he cannot see whatever he would like to be able to see but only what there is to be seen. He can see for instance that the upper leaves of the tree are smaller than the lower leaves; but he cannot see that the upper leaves have smaller cells than the lower leaves. Novelty requires that he surrender to the insistence of what lies before him and what therefore he cannot deny.

Thus it often happens that simple observations are matters of chance. Malus' discovery in 1808 of the polarization of the reflected light resulted from an unexpected observation. When Galvani in 1786 accidentally touched the leg of a frog with an electric wire he did not know that the leg would twitch. At the level of ordinary common sense, observation can happen by chance without any aid from instruments or from mathematical calculations. Nicolle's observations of the transmission of typhus by fleas is a case in point. Nicolle observed that typhus patients were no longer contagious once their own clothes had been removed and they had been cleaned, and he correctly inferred that the fleas in their clothing had been carriers.

The scientist believes in the existence of the material object independent of his observations, otherwise he would not observe. He would not observe, for instance, if he thought for a moment that he was limited to the observation of his own mental states. Science does not in fact take off from high abstractions but from sensations. Empiricism has subjective roots. But to observe is somehow to read the events recorded in the nervous system as reports of an external world. In the last analysis, the empiricist says, I will believe what I perceive by means of the senses, and I will let myself be led to believe nothing else; hence any beliefs which are to be accorded the adjective, scientific, must have to do with events which in some aspect have been hooked up with the nervous system. This is the aspect of science which so impressed the positivists. There are no doubt "protocol sentences" and "atomic propositions." The only question is, are these anything more than the descriptions of initial stages in the investigation? The matter cannot be allowed to end there if we are to admit the claim that science has supported. For the investigations of science are not after all those of the nervous system of the observer – science is anything but introspective – but of the world which the stimulations of the nervous system report.

Raw data are presumed to underlie all of the interpretations, constructions and inferences which are put upon them. The effort to penetrate to them is the most urgent demand of empiricism, one upon which the whole import of the scientific method of investigation is founded. For it is not the privacy of "sense data" which the scientist is seeking but rather those objective and independently based properties which the impressions obtained by the senses disclose. Simple observation involves the acceptance of the material object just as it appears, and involves the dependence of the observer upon it. He is aware of it, and at the same time powerless to make it appear other than it does. The object is genuine and makes its impressions upon him. One of the assumptions upon which science relies is that there are aspects of objects corresponding to the stimuli of the nervous system, that somehow sensations can be read as messages received from a world external to the senses, which it is the business of the senses to record.

A strong piece of evidence for the objective existence of the source of the stimuli received by the senses comes from the senses themselves, not considered singly but taken together. The coordination of particular sense experiences would not be possible were there no material object; for it is in the object, and not in the subject, that the report of the senses can be unified. This piece of silver the observer sees as a grey metal he can also touch and to some extent manipulate; he can hear it drop with a clangor, and, if he likes, note its metallic taste. Were there not sufficient evidence for the objective existence of what a single sense discloses, as some radical empiricists have contended, at least the same statement cannot be asserted for the disclosures of the coordinated senses, which can be brought together only to the extent to which they can be focused on something external. It might be possible to confuse a material object with the single perception of that object, a tree with the sight of the tree; but such a confusion is clarified when we add to our sight of the tree the additional sights from other perspectives, and also the feel, the smell and the sound of it. These experiences taken together with some kind of persistent relation constitute what we come to recognize as the object, tree.

Empiricists, at least early empiricists like Mach and Avenarius, are caught in the subject-object dilemma. There is indeed much here to trouble us, and it is necessary to remember only what it is what we take for granted. There is for scientific purpose no absolute evidence for an external material world, none, at any rate, that will stand examination. The existence of the world is assumed, and the deductions from such an assumption have justified it. For there is no demonstrable solid ground under empiricism. However, on any other assumption, further investigation would be rendered difficult. What could we investigate, and how could we proceed either in theory or in practice if we were to assume that nothing lay beyond the nervous system? We should soon come to the end of our inquiry. And, strangely enough, we should probably in that case know less than we know now about the nervous system itself. Thus the existence of an external world independent of all investigations is a requirement of the scientific method.

Science has always been realistic in assuming that a material world exists to be known, and has assigned the sense organs the lead in the effort at knowing it. Hence the role of the senses in observation goes along with a belief in an observable world. "Observable" means that there is something which can be observed, something objective. The existence of a subject implies not subjectivity but the presence of a real object, and there would be no use for an investigator were there nothing to investigate. Could science posit no objective reality for its concepts, as Clifford and Poincaré, for instance, asserted, science would be a description of our experiences and nothing else, thus making physics and chemistry into branches of psychology. This position overlooks the fact that every observer does locate some things in the internal world of the subject and other things in the external world of objects, a distinction he needs in order to account for action and affection. Mach was too restrictive in endeavoring to prevent the elaboration of scientific findings. He wanted science to end with observations conducted by means of the senses, whereas that is merely where science takes its start. He was correct only in asserting that scientific formulations must somehow eventually be referable to a world that can be disclosed to the senses and failed to see that this very requirement implied the existence of an external world. Roughly speaking, scientists endeavor to rest their findings on the observation of fact. But there are difficulties in the way of establishing formal rules for such procedures.

In the first place there is the variable comprised by the observer himself. No one sees exactly the same thing on two separate occasions, and this is true not only because the object does not remain the same but because the observer does not, either. Small errors creep into observation due to a thousand small changes in the condition of the observer: large changes of which he may be aware or for which he may attempt more or less successfully to compensate, and very small changes of which he may remain unaware but which may nevertheless account for errors of observation. For nobody will deny that there can be observational errors.

In the second place, there is the variable which is the observed object. The attempt to establish an infallible criterion of what constitutes observability has failed. What is observable today may become unobservable tomorrow; the large herds of wild animals that can be observed ranging the grasslands of Tanganyika will probably not be alive in a few more decades. Conversely, what is unobservable today may become observable tomorrow. It was once thought that it could not be decided whether there are craters on the other side of the moon because the other side of the moon could not be observed. That issue was first raised in 1936, but in 1959 the other side of the moon was photographed. The contention that science rests on the observation of data should be revised to read that science rests on the observation of data potentially observable, with the provision that the range of the potentially observable is a shifting one.

Considerable training on the part of the observer, as a matter of fact, is necessary. In the above paragraphs we have been assuming that an absolutely accurate instance of unaided observation is possible. Yet the fallibility of perception is a notorious fact. It is a favorite topic in courses on elementary psychology. It is an even more painful fact in courts of law. The variation between the evidence given by witnesses to the same events is a phenomenon which renders the findings of the facts an affair of great difficulty if not an impossible one. Obviously, with small corrections of spatial differences in perspective, the witnesses to an event have all seen the same event. They report it quite differently, however. The reason for this is that they have highly diverse sets of experiences in terms of which their seeing is done. They have, in other words, different hypotheses. We are assuming here an honest set of witnesses. No lying is done, and each has reported exactly what he thinks he "saw."

Training in accuracy of observation is of course possible. To learn to detect what is before one's senses while holding to a minimum the element of interpretation which arises from previous experiences and beliefs and other such determinants, is no easy task; but its accomplishment can be improved with practice. This is indeed one of the things the apprentice does learn in the laboratory and the field. It is no small part of the equipment of the professional scientist.

The first step in any science for a trained man, then, is observation. To perceive what appears with as little interference as possible from the perceptual process, skill is required. Existence as a whole consists in a complex of undiscriminated occurrences. To sort out some for special attention as facts is to have presumed upon the special knowledge acquired in the training. In an electron micrograph of a cell from the pancreas, for example, only a highly trained observer would be able to distinguish the nucleus, the sub-microsomes attached to the microsomal membranes, and the various shapes of mitochondria. He would have to know in the first place that such entities existed and what they were like. And his training itself, of course, would have relied upon the deductive system of the science in which such training took place. Hence when we attempt to discriminate any absolute starting point in the scientific method, we find ourselves enmeshed in a regress. The collection of facts, the ascertainment of data, is made by means of a certain amount of information, acknowledged or unacknowledged. Thus science does not operate merely with the facts observed but rather also with the information which makes the observations possible.

We shall have to assume that observation as such is not scientific; science is a matter of *trained* observation; systematic, planned observation, disciplined observation, schooled to omit nothing relevant yet to observe what is of significance and what can be repeated, in a word, sophisticated. Repeatable observations belong to the domain of the scientific. The method of science is self-corrective; the deliberate planning of successive repetitions furnishes the mechanism of this self-corrective process.

Experience to be called scientific must be repeatable. There are plenty of those who are ready to tell us of their unusual experiences, and if we admit on subjective grounds what has been experienced, then revelations and all sorts of super-natural phenomena become ostensibly amenable to empiricism. In scientific observation, the data of experience must be held down to what can be repeated, and which therefore must be objective since it can be social. It is not only the particular experimenter who can repeat his experiences, but anyone; and if his reports of his observations have been correct, his colleagues will obtain the same results. The purpose of experience in scientific method is to disclose to us the data corresponding to the experience. In other words, observations must be objectively interpreted if they are to have any scientific worth at all.

There has been in the world long before the advent of science a very old profession in which observation is paramount, and this is the profession of the artist. And so it may be helpful to ask, what is the difference between the observations of the scientist and those of the artist?

First of all, the similarities will have to be freely acknowledged. The scientist is perhaps more of an artist than he has cared to admit. In painstaking preparation, in care of training, in meticulousness of observation, the scientist, like the artist, takes readings of the external world. How far from the scientist was a painter like Cézanne when he described his method as sensations in the presence of nature? Yet science, we will have to confess, is not art; wherein does the distinction lie? There are two important differences between science and art, and since they show something of the nature of the scientific method by contrast, it will be helpful to mention them.

The observations of the scientist are repeatable by others, whereas those of the artist are not. Any number of scientists will obtain the same readings from the same observations. After the results reached by Michelson-Morley with their interferometer, a number of such instruments were constructed by other physicists, and the reports for a long while did not differ substantially from that of the original investigators. It is clear, however, that no two artists see thinks in exactly the same way. Two painters working from the same landscape will produce strikingly different pictures.

To the contentions in this argument, several counter-arguments could be made. The first of these is that the observations of the scientist are not always exactly repeatable. For example, the repetitions of the Michelson-Morley experiments with the interferometer to detect the ether drift made by D. C. Miller and his associates over the years 1902 to 1926 discovered a small positive effect for the ether drift, yet the theory of relativity was not abandoned. But the answer to this is that whether relativity was founded on this experiment or not (and Polanyi, for instance, thinks that it was not) there are other confirmatory reasons for accepting it that collectively seem stronger than this one piece of evidence against it.

The second counter-argument is directed against the assertion that the observations of the artists are not repeatable, and it starts from the phenomena of fashions in art. Many artists evidently *do* see things in more or less the same way; otherwise there would be no art movements, such as impressionism, cubism, or abstractionism. But here the reply is that fashions in art which are so strong as to obliterate individual differences occur only among mediocre artists. The great artists are those with originality of vision and insight.

Another distinction between science and art is to be sought in a difference in subject-matter. It lies, presumably, in the disparate kinds of elements in the external world which the scientist and the artist seek in their observations. For the scientist is looking for laws: structured invariants, while the artist is searching for material symbols of values. Science, it has often been noticed, is value-free, while art is almost entirely a matter of values. These values may be hung upon a complex frame or structure of relations, as for instance is the case with polyphonic music; yet it is the values and not their underlying framework with which the artist is in the end concerned. The scientist, on the other hand, undertakes his observations in order to determine just what relations exist. Powerful forces – values – may be and usually are involved in the subject-matter with which the scientist treats. Yet it is not the values *qua* values nor the forces *qua* forces with which as a scientist he is immediately concerned. Atomic power may be represented in equations and the molecular forces in a biochemical formula set forth in diagrams. Such forces and values have corresponding abstract structures, and he is concerned with the structures and not with the values which emerge or with the interplay of forces, though he must cope with these, also. He seeks causal laws, statistical frequencies, regularity of events; his aim is mathematical formulation, and so these formulas reveal themselves to him only to the extent to which they can be rendered precise. Thus the reasons why he is observing must determine to a large extent the results of his observations. If he is an honest observer, maintaining the proper sort of detachment, he will not determine in advance what it is that he is to see, but will decide in terms of his interest what *kind* of thing. He is looking usually for the regularities representing some selected segment of nature.

Observation, then, is not altogether simple; it is colored by what the observer has already experienced. It is made from some perspective which is at least partly peculiar to the observer and his interests, and a little determined by what he expects to find. Selection and classification are inevitably involved in every act of observation. This does not mean, of course, that scientists do not hold up to themselves the ideal of perfect detachment and impartiality; but it does mean that such an ideal can only be approached, not reached.

Observation, then, is a rudimentary kind of interpretation. To what extent can it be random, if this is true? The scientist selects what he plans to observe and maintains at all times control over his observations. Facts nevertheless are objective, and they resist his will to see them altogether as he would, else there would be no disappointing results in planned observations. The attempt to ascertain what the facts are has to reckon with the elements of fallaciousness peculiar to the mechanism of observation. The faults need not be fatal. For colored and selected, distorted even, as all observations may be, there is none the less a core of genuine fact in them.

The effort to get at this core encounters serious limits, for what makes it possible also prescribes a set of boundary conditions. There are pre-observational assumptions which are difficult to isolate; and there are still other obstacles. For instance Jevons has suggested that if we knew how much bias enters into observation we might take it regularly into account and so improve the authenticity of the report of natural events, and this discrepancy discloses a deep activity of selection; the scientist actively chooses what it is that he shall passively observe only after some casual and perhaps unnoticed orientations have formed in his point of view. No doubt the recall of information plays a role, so that memory cannot be excluded as a factor; but there are also the elements of accident and surprise. His expectations are continually being overturned by the data of experience, because they are framed in terms of his previous knowledge; and however much that knowledge is derived from experience, it is bound to be limited: quite obviously he expects that there is more than he has experienced, for otherwise he would not observe, but paradoxically he expects it to be familiar.

But almost immediately we shall have to go back of these statements in order to discover what they have involved us in admitting. For the pure phenomena are linked by the boundary conditions to a theory of perspectives. The observer perceives what he is determined to perceive by the object, but not all of the object in any one observation; and what aspect of the object is conditioned by the perspective occupied, and perspectives in turn are determined by the training and equipment, the ability and peculiar interests, of the observer.

Simple observation is mediate knowledge in that it is acquired by means of a perspective, and immediate knowledge is that it is determined by what is available from that perspective. It cannot be said to be free of all inference but then neither can it be said to be determined by inference. As grounded in conditions which are tacitly established and which determine both the passivity of the subject and the character of what is to be observed, the silent assumptions concerning the aim of the observation, simple observation is mediated knowledge. But within the limits set by the perspective from which the observation is made (and these narrow rather than distort the observation), simple observation is immediate knowledge.

It may be recalled that the aim of simple observation is to record what happens under the assumptions of pure phenomenology which is the effort to become passive in the face of the material object. But it must not be supposed that the result is a distillation of pure phenomena devoid of any interpretation whatsoever. It is important to remember that it is not the entire external world that is placed under observation by a given operation in science but always some particular segment of it. And how is this segment chosen? We do not observe entirely at random but usually in some decided fashion. And the selection of the segment of the external world which is to be observed is made in terms of some previously adopted hypothesis assumed or explicit. A ghosted hypothesis guides the selection of the data. The hypothesis in this case may be brief and tentative, but it is there implicity supporting the choice of the area of observation.

There is no such thing as seeing unaided by previous theory. We must distinguish between the intrinsic quality of what we see and the extrinsic meaning. There are obstacles in the make up of empirical data: how reliable is the account of what was observed? How can the investigator be sure that his facts are reliable enough to be regarded as data? There is room for errors of observation which the process itself must expect to encounter. In simple observation there can be no error – the impressions are whatever they are. But in the description of the simplest observation errors can creep in, for description requires at least rudimentary interpretation. The observer thinks that he "sees" a sunrise but he does not. What he actually sees is a glow at the center of his field of vision, a patch of color. But he read it as a movement of the sun, and he knows now that it could with equal justification have been called a horizonfall. Perceptual errors are too well known to require illustration; illusions and hallucinations, for instance.

The point is that in simple observation we have already discriminated among the observed material in terms of the implicit hypothesis adopted. Observation is the simplest kind of abstraction, a sort of rudimentary induction. And the abstraction is worthless unless we can connect it with others under some comprehensive theory. Otherwise we should have to ask what the observations had been for. We haven been talking of course about observation without instruments. Science dispenses with such observations very early, and calls upon the aid of instruments. Instrumental observation is a kind of primitive experiment, concluded with pointer readings. And pointer readings must then be interpreted by means of some abstract theoretical system. The scientist is quite able to explain his ideas in elementary terms, only, these, too, are abstract. A science is a system of relationships. The search for laws is derived from the observation of the behavior of similar material objects. Without an assumed background, all explanation sounds complex. But complexity is a relative affair and a matter of degree.

2. *Controlled observation*

We have now to discuss controlled observation. In general this step consists in an elaboration of simple observation, and makes way for later stages in the scientific method. We have noted that observation of the scientific sort requires self-conscious naïveté, the ability to perceive as innocently as possible what lies in that portion of the external world which has been selected for consideration. A further development involves another set of properties. In controlled observation, training in motor responses and manipulative skills are needed. Feeling and action are dominant over thought. Of course, it is to be presumed that when we talk about the beginning of a scientific procedure, we are doing so in terms of a professional scientist, and a professional scientist is a man with both a strong deductive background and some active experience in his chosen science. Science as a rule does not begin with random observation and it certainly does not begin with experiment. It begins with the controlled observation of facts and with the problems to which these give rise. Observations are not random; the scientist selects what it is that he wishes to observe. Some data are so obvious that a trained observer would find difficulty in avoiding them, while others have to be trapped by means of intricately devised instruments. In any case, controlling the conditions under which observation is to take place means also to some extent controlling the observations which are to be made as the result of it. By controlled observation is meant the procedure of deciding by means of action the conditions under which simple observation is to operate.

There are three separate and distinct ways in which observations can be controlled, and all of them involve the making of preparations. These are: (a) adjusting the object in such a way that it can be brought into focus, (b) getting the ground ready for observation, and (c) selecting the conditions under which observation is to take place.

(*a*) The first variety of controlled observation is the one calling for the handling of the material which is to be observed. The proper isolation of the object is a crucial prerequisite to successful observation, and this requires of course the ability of discriminate between true and false isolates. Extensive labors consisting in operations performed upon the subject-matter will often be required before the investigator can begin his observations. The preparation of microscopic slides is perhaps the best example of this variety of controlled observation, but there are many others: the making of thin sections, for instance, the immersion in a fluid medium, or the growth of a proper bacterial culture. Much care must be taken at this stage if the observations which have been planned are to produce the most satisfactory results.

Controlled observation differs from simple observation in that controlled observation is guided by rudimentary hypotheses, adopted or assumed. Here the hypothesis begins to come more to the fore than it did in simple observation. Observation is of individuals, while hypotheses are general presuppositions. Observation of the controlled variety can be analyzed into sensation plus deliberate interpretation, and deliberate interpretation requires hypotheses which are more or less formed. The interpretation at this stage is complex. Good observation requires active speculation, for which theory is needed. The cell theory of Schleiden and Schwann has made it possible for trained microscopists to observe cell division. No one has ever observed hungry rats, or the death of a man; but the efforts of the rats to get at the foodbox, and the sudden transition in the man from activity to a lack of all response to stimuli, are so interpreted. To engage in deliberate observation under controlled conditions means in each case to draw heavily upon hypotheses guiding interpretation from the vantage point of presuppositions. Why would the observer in the laboratory prepare slides for microscopic investigation, the observer in the field make elaborate provisions for not disturbing the nesting birds, or the builder of the telescope hope to see astronomical objects on a larger scale and in a sharper focus, if he were not operating in terms of important assumptions? And is it not appropriate to regard such assumptions as the silent but forceful employment of hypotheses?

As a rule, what is available to simple observation are the middlesized objects disclosed to ordinary experience, the furniture of the mesocosm. What we might normally encounter in our daily lives without recourse to special efforts at observation or to special instruments is what simple observation works over. But in controlled observation what is usually observed is what is *not* available to ordinary experience. Controlled observation generally penetrates both below and above the ordinary levels of analysis, to entities and processes of the microcosm and macrocosm respectively. The result is that there is a change in the character of the facts observed. In the more advanced sciences, the gap tends to widen between the experience of the enlightened but unaided senses and the kind of information that is made possible by instruments. The observation of sub-atomic particles indirectly by means of tracks in bubble chambers; the observation of astronomical events directly by means of the telescope, or indirectly by means of instruments attached to the telescope or used in connection with it, such as light filters, the camera or the spectroscope; these are cases in point.

Observation itself, of course, takes place in order to serve the search for invariants. Thus controlled observation tends to be both prolonged and repeated. An investigator may spend many long hours and even many days at the eyepiece of a binocular microscope or of a reflecting telescope, or before a cage containing experimental animals. Leavitt's extensive observation of Cepheid variable stars in the two Magellanic Clouds led to her discovery of the period-luminosity relationship: the longer the period of the star, the greater its absolute magnitude. Prolonged observation requires practice in concentration, and this in turn presupposes a sophisticated kind of training.

(*b*) The first variety of controlled observation involved the preparation of the object. The second variety depends upon an arrangement of the relationships between the observer and the object. Tinbergen's careful arrangements for observing the herring gull colonies in their natural habitat: the setting up of a blind as unobstrusively as possible at the best distance from the colonies, and the patient awaiting of the proper occasions for observation, show the lengths to which it is often necessary to go to make observations meaningful.

There is always an element of anticipation in controlled observation: the investigator has decided in advance what kind of thing it is that he wishes to look for, and such anticipation is framed in terms of an hypothesis. For instance, when radio telescopes were turned in the direction of the constellation of Cygnus, more specifically on a radio object known as Cygnus-A, a very powerful emission was recorded. Further resolution by means of optical telescopes revealed an object which has been interpreted as an instance of colliding galaxies. The use of the optical telescope was an effort to confirm or disallow the hypothesis, suggested by earlier work with the radio telescope, that there was an astronomical object of more than routine interest.

It becomes increasingly obvious that controlled observation is the same as preliminary experimentation. The human senses alone are hardly sufficient to deal with the recondite areas of nature which the scientist seeks to observe. Thus instruments play a steadily greater role. For instance, movements can be observed within plants by the use of time-lapse motion pictures. Without instruments, the sense organs could not be extended to the perception of the fine structures with which science is now familiar. Most scientific observations take place beyond ordinary common sense levels and are accomplished by means of instruments. For instance, in 1936 Zwicky, using an 18-inch Schmidt-type camera at the observatory on Mt. Palomar, conducted a systematic search for supernovae, as a result of the discovery of the first one in 1885. The construction of instruments is part of the observational technique, and often very delicate apparatus is involved. Much skill is called on for both the construction and the use of scientific instruments. These make observations possible that would not be conceivable without them, but they also impose their own limitations, which must be recognized.

In the end, however, it is the object observed which governs. Indeed, such is the whole purpose of observation: to discover what the investigator has to learn from the object, as opposed to deciding about it in advance; to learn, in other words, what it is that he will have to admit. No matter what pains are taken to isolate the field of observation and to sharpen his ability to observe, it is still necessary for him to be prepared to concede what is observed. Even in the case of controlled observation with its assumed hypotheses, with its anticipations and its instrumental aids, the tyranny of the material object prevails. F. Sanger spent ten years at Cambridge University studying the structure of the insulin molecule. Observation, then, facilitates the cognition of the object, and, more often, of the special features of the object, such as its properties or its behavior. These constitute the facts as observed.

(*c*) The third variety of controlled observation requires a selection of the conditions under which the observations are to take place. Galileo's construction of the first telescope was just such a preparation, and so is the invention or employment of any device useful to observation. For the observation may involve an extension of the range of the ordinary senses, as with the telescope or microscope, or it may involve indirect means, as with the vibrating membrane of the Geiger counter. A rather elaborate example of controlled observation is the greenhouse laboratory constructed at the California Institute of Technology by Professor F. W. Went and his associates in which plants can be observed under ideal circumstances; for the ventilation, the contents of the atmosphere and its temperature can be rigidly maintained.

But perhaps the most radical advance in the technique of observation involves the freeing of the laboratory from its fixed position. Hitherto it had been necessary for the scientist to bring his specimens into the laboratory; now it is rapidly becoming possible to bring the laboratory to the specimens. Observational instruments are being sent into orbit, they have been sent to the moon and to fly by Mars. The observer no longer has to be personally present at the scene of his observations. Reports of them can be sent back to him.

A very important asset of the observations peculiar to science is, as have noted earlier in connection with simple observation, the possibility of repetition. The same observations can be repeated (at least approximately) by the same observer on different occasions, or by different observers. Repeatability of observation gives greater assurance in science of what has been observed, and no scientist would object to another scientist's checking on his report of observations. Every observer in science is so to speak on patrol and alert to the importance of what he may note for the first time; but his observations must be corroborated, and there are very few exceptions to the need for the support which is furnished by the confirmation by others who have made the same observations after him. Now, the first observations of a set may be simple observations in this sense (though in this sense only); but consequent repetitions are controlled observations: they will have to be determine in the ways we have enumerated: by manipulating the material to be observed, by arranging the relations between the observer and the object or by selecting the conditions for observation.

In one sense the repetition of a simple observation *is* a controlled observation, since the second and all subsequent observations have to be managed at least to the point that they are planned. The scientist is interested in permanent features of the natural world, and one indication of their permanence is that they can be approached again and again with approximately the same results. Patient observations may be repeated; they may be made by the same observer over and over, or by different observers on different occasions. Repeating an observation is in a certain sense an intensification of it; but there is still another way in which controlled observation operates. This is in varying the perspective on the object observed. A material object or event possesses features not available to any single perspective; thus varying the perspective is a way of collecting observations on the same object. The systematic variation of perspectives is a way of controlling observations available to the investigator as one of his stock of resources.

Chance often operates to preempt the scientist's attention, but more often he has planned to devote to it a certain set of manipulations. The unexpected results of manipulations in a laboratory or of planned conditions in the field are equally controlled observations. The greater the degree of preparation, the more likely that the observations will be undertaken at high analytical levels. No one ever looked through a microscope or a telescope to observe events which are available to ordinary observation. The astronomer knows that if he takes time exposure photographs through telescopes, stars visible neither to the naked eye nor, for that matter, through the telescope itself, will appear.

There is another kind of observation which has always to be included under controlled observation, and that is indirect observation. Some instruments are used to aid observation directly; the telescope or the microscope for example. But others aid observation indirectly, such as the centrifuge or the cyclotron. The more a science progresses, the more indirect observation takes the place of direct observation. In indirect observation, something observed directly is read as evidence for something else not observed: tracks in the cloud chamber as evidence of the formation of ions; overt human behavior as pathological evidence of the deeper existence of psychoses; a chemical reaction taking place in the presence of a substance, unaffected by the reaction but non-existent without it, as evidence for catalysis.

Perhaps the most common kind of indirect observation is that involving pointer-readings. Such observation at second remove consists in, in addition to the reading of dials, the interpretation of photographic plates and a host of similar devices. The observer will be called upon also, for instance, to count flashes on screens and clicks of counters. As scientific investigation penetrates to deeper and deeper levels of analysis, it relies more and more upon such indirect means of observation. Whenever the range of human senses is to be extended, this must be done in terms of instruments which do our sensing for us. And even within the human range, to obtain the kind of accuracy of observation required by the scientific investigator, instruments are more suitable than human judgment. A thermometer gives a more accurate account of bodily temperature or of the atmosphere than do the fallible human senses. Assuredly, not all instruments serve observation. Many are employed to perform operations which could not be accomplished in other ways, such as the ultracentrifuge; but from the cathode ray oscilloscope to the hydrometer, from the radiation pyrometer to the stroboscope, indirect observation and hence controlled observation prevails.

Instruments make observations more complex. As we complicate the investigations we also increase the opportunities for making mistakes. We have seen that in the case of simple observation error could mean only that we were not reading the observation properly. In controlled observation we have to some extent the possibility of increased interpretation and hence of increased error. For now there are instrumentally produced errors to be added to the ordinary errors in interpretation. Always the observer is under the necessity of referring his observations to other observers for purposes of repetition. Repetition reduces (though it does not necessarily exclude) the probability of error in some kinds of investigation, notably, in those kinds which do not involve very large, complex and advanced hypotheses, where the degree of accuracy warranted by the experiment is considerably exceeded. The observer may have been at fault or his instruments.

We have already noted that controlled observation gets away from what is available to ordinary experience at the common sense level. In selecting the conditions for observation and in the further selection involved in what stands out from prolonged observing sessions, scientific investigation has got still further away from the casual and accidental observations of ordinary experience. The observations of the trained investigator are usually deliberate, and do not take place merely because he happens to be alive and to have the requisite number of functioning sense organs. With even the most primitive stage in abstraction, a certain degree of stability results. The observer begins to think in terms of data which are independent of his observations and which can as a consequence be retained.

3. *Observed facts*

We began this chapter with a discussion of simple observation, and then moved to consider observations of a more sophisticated sort. In both cases, of course, we were concerned with the kind of observation with which science starts. We have reached the point in the argument where we shall have to take formal cognizance of what is observed. It is difficult enough to determine precisely what is observed. But we shall still be talking about the way in which observations are made and concepts are formed in the laboratory and in the field, as distinguished, for instance, from mathematical and linguistic practices. We have been speaking of the observation of fact from considerations of observation; now we shall look at the same procedure only this time in connection with the results of observation: from considerations of fact: To acknowledge that the material object observed has about it the character of fact is to make the first rudimentary kind of interpretation. In this sense, thought is dominant over feeling and action. In all three steps, namely, in simple observation, controlled observation, and observed fact, we are gradually separating the object from the observer in order to consider it in isolation.

(*a*) *Varieties of data.* A datum is a disclosure of experience, interpreted as some aspects of a material object or event whose existence is held to be irrefutable. An observed fact is the report of a datum, or a proposition in which a datum is asserted.

Thus a datum can be either a non-repeatable individual or an empirical theory whose unexceptional nature seems equally well established. In either case, however, the observer is confronted with something which has both a particular and a universal aspect. Particulars can be sensed, yet they are apparently lacking in analytical limits: below any given level of analysis, there are other levels; for instance, below the atomic level lies the subatomic, and below the subatomic other levels are suspected. The world from which scientific concepts are to be derived by means of observation is one containing irregularities as well as regularities; universals such as qualities, quantities and structures, and also unfathomable individuals whose susceptibility to analysis appears unending.

The particular illustrates the universal. Thus if the fact chosen by the observer be a combination of boron with oxygen, he can see this happening at 100°C. as a brown amorphous powder burning with a green flame. That this can be done again and again indefinitely points to the existence of a set of general conditions under which classes of substances react uniformly; universals galore but combined uniquely on each occasion. Whenever a universal principle is illustrated it is a particular material object which illustrates it.

That there can be no contradictory facts shows the way in which laws are derived from the disclosures of experience. Magoun has pointed out that the results of observations which seem inconsistent are often the source of unexpected knowledge. It is helpful to compare the impossibility of inconsistent facts with the possibility of inconsistent propositions. The observed facts point toward laws rather than toward other facts.

The observed fact is, of course, an abstraction: boron and oxygen have other properties, and there are other amorphous powders and green flames. Any one relation between two elements is simpler than the two elements taken in their entirety and including all their relations; otherwise there would be no abstractions possible and hence no facts ascertainable. That the United States is north of Mexico and that water freezes at 32°F. are data. A very considerable amount of background interpretation is assumed in each instance: the concept of nations and their relative geographical position on a projection, in the first case; and in the second the character of a fluid, the temperature measurements on the Fahrenheit scale, and the qualitative changes involved in freezing. Despite the difference between the two cases, the first so general as to require the specification of extenuating circumstances, such as mineral content, pressure, altitude, etc.; the second specific, each may legitimately be called a datum. And it is characteristic of data that they resist all efforts to make them other than they are. Data may be variously interpreted but they cannot be changed. It may often be difficult to ascertain what are the facts in any given case; nevertheless they are the facts.

There are, roughly speaking, five varieties of observed fact and we shall enumerate them here; although no doubt, finer discriminations could be found.

Simple sense perception guided by contrast discloses the barest fact. The biologist observes through a microscope that this thread is no longer than that thread, that this globule is darker than that globule. Thus his interpretation of mitochondria, for example, is held to a minimum, though some degree of reading is unavoidable. He interprets when he singles out from the sensory background certain elements and their relations, but he strives to do so no more than is necessary to make the observation.

At the next level of fact the observer discriminates between substance and properties. This lead is a good electric conductor, that piece of carbon-saturated iron melts at 885° Centrigrade. Here interpretation has been added quite frankly to the findings of sense experience, and the meaning of the experience has been emphasized. Suffice to say that no one has ever "seen" a substance or a property. What has been seen has been interpreted in this fashion on a basis which is considered to be sufficient evidence. Already at this level hypotheses can be framed, we are sufficiently removed from ordinary sense experience for the occurrence of error to be possible. What makes lead an electric conductor, what delays the melting of iron until this relatively high degree of heat?

We now move to the next level where the observation of facts turns on complex functions. For example, water electrolyzes into hydrogen and oxygen in proportions of two to one; again, when the temperature of a gas is constant, the volume is inversely proportional to the pressure (Boyle's Law). Experiments can be performed in which these relations can be "observed," that is to say, given a trained observer, what is observed can be interpreted only in this way. Suffice to add that by now simple observation has been left far behind and we are in the area of controlled observation. We are in the presence of the invariantive associations of materials or of their behavior (events). If the evidence produced by observation seems to the observer to be sufficient, then he asserts it as a fact. There is a borderline here between fact and theory which has to be decided in each case. That *all* men are mortal, is a fact to those observers who have seen men die and who have scanned insurance actuary tables, but not to those who believe in miracles. The scientific observer comes to his tasks with special assumptions as part of his equipment.

Another level of observed fact contains the reactions which can only be observed indirectly. When a deuteron is formed from a proton and a neutron, a photon is released with the binding energy of 2·2 Mev. Proteins are hydrolyzed by the proteases in the human digestive tract into alpha animo acids. The "observer" here is elaborating findings based on inferences from observation; he has reached the level of mathematically-expressed relations supported by experimentally-discovered data. At such complex levels of analysis what are called data are the interpretations of indirect observations by means of mathematical structures.

The last level of observed fact is that of the entities and processes postulated in order to render a causal account of phenomena. Dirac supposed the existence of an entity: the positron, and a process: annihilation radiation, before they were confirmed by experiment. Often what is observed cannot be explained except on the theory that such entities and processes exist. Experimental results may be of this character, as was the case with Dalton's attempt in 1803 to explain the properties of gases on the basis of the existence of extremely small particles or atoms. Freud's explanation of psychopathology by means of the interaction of the ego, super-ego and is also a case in point. Hypothetical entities and processes if sufficiently well-established are counted as data.

The criterion of classification employed here in describing the varieties of data has been that of increasing abstraction. The justification for the selection of such a criterion is to be found in the direction of scientific investigation. Data are basic, but it is theory which is sought.

The verification of a theory depends upon its support or allowance by data. In some tentative areas of study, at the beginnings of any empirical field, the data adduced are not final but merely exploratory. The range of exploratory facts is where the speculative and the empirical meet. For the search for facts already assumes some interpretation. Accidents apart, it is difficult to see how data could be searched for without a theory of some sort. Even though the data govern, they are not alone, else theory would be out of place not only in the search for data but in the connection between data.

The borderline between data and non-data is often very dim. What was a datum yesterday may not be considered one tomorrow. The existence of irrefutability in the class of occurrences based on observation is often difficult to establish or to maintain. The "aether" of nineteenth century physics, which was supposed to pervade all space and matter and act as the medium of the propagation of light, is no longer accepted. Because facts are sometimes long-established principles concerning particulars, the appeal to a fact may often mean no more than reference to a long-established principle.

What was a theory yesterday may be a "fact" today. To enlightened common sense, gravitation is no longer a theory but a fact, and many such facts are simply familiar theories. Similarly, there have been concepts which have knocked on the door and begged for admittance as facts, and some have succeeded while others have not. The concept of "field" in electromagnetics succeeded; "phlogiston" did not. The observation of fact is a process which corrects itself from time to time, and is not absolutely rigid about what is and what is not observed.

(*b*) *The character of scientific data.* Data, then, are not just anything observed unless, that is, we take observation to involve some kind of discrimination. For no one observes the whole world but only some part of it which has been selected for that purpose. The problem of analyzing observation is complicated by the fact that in every object or event there is more observable than lies within the power of any one observer to detect. Existence is everywhere dense, and it is only the untrained observer who would doubt the force of the difficulty. Psychological experiments have demonstrated time and again, and the law courts are painfully familiar with the fact, that no two observers report the same occurrence. Observation for scientific purposes is a kind of rudimentary experiment. We perceive only by selecting to some extent in the very act of perception. Simple observation uncontrolled and unconditioned would seem to be an ideal of science to which it never altogether attains and toward which it strives only in its earliest stages.

In changing over from the point of view of the subject implied in observation to the point of view of the object implied in fact, the multiplication of the senses calls the turn. Even simple sensory observations does not depend upon the isolated findings of a single sense; instead, a number of senses are brought to bear upon the same material object, and they can be coordinated only on the basis of an assumed objectivity for that which is observed. Hence full observation leads eventually to objective fact.

What confronts the observer is usually a choice of facts. Events have a way of outstripping observations, and there is a richness to existence that compels a selection. A datum by itself is nothing, and "the point of view of fact" is preserved only by that perspective of the system in which the fact is imbedded, which is available from the fact. The selection is made in terms of provocative fact. By "provocative fact" we shall mean data not explained or accounted for on the basis of accepted theories and therefore stimulating to further inquiry, something close to what Bacon called "prerogative instance," or what Herschel called "clandestine instance." From a scientific point of view, the most satisfactory result of a provocative fact is the discovery of an hypothesis. It leads to the making of more observations, suggests new speculations or even new theories. It is important in one of two ways: because it is peculiar in some way which seems significant, or because it is typical of its class. It is a conductor, not a terminal point of investigation.

Scientific observation begins at the level of ordinary experience but does not remain there. To attend to some portion of the undisturbed environment with the naked eye or ear has its place in science, but it is an elementary place. Most scientific observations, as we have already noted, are made of an environment which is conditioned and by means of instruments to aid the simple senses. Crude facts become scientific data when observations take place in some indirect fashion and at levels of analysis which lie below those of ordinary experience. The investigator watches pointer-readings or examines photographic plates obtained by using microscope or telescope – all observations in the end submitted to the sense organs yet enormously conditioned in the process of preparation, and reporting a world which for the most part lies well beyond their range.

Observations may, in short, be referred back to common sense, and their planning may take off from the level of common sense, but they go through a crucial and significant phase where they are far removed from common sense. Scientific fact is obtained at high analytical levels. There the observer has left ordinary experience far behind and often indeed encounters facts which contradict it. Familiar materials, such as water, metals and rubber when subjected to high altitudes, heavy pressures or enormous speeds, behave in quite unexpected and surprising ways for which ordinary experience had never prepared the observer. Corresponding to the element of anticipation in controlled observation, there is the elements of surprise. It often happens that careful arrangements are made to undertake certain observations when what is observed is not what was expected.

What we shall have to acknowledge, in other words, is that science involves a reference to *reason* and *data*, not to *ordinary experience*. It is an error to identify reason with ordinary experience, for there are levels of inquiry which reveal facts contrary to ordinary experience. It is not common sense to think that diamonds can be manufactured, that light rays can be bent, that "nature abhors a vacuum only up to 35 feet of water," or that matter in an excited state is more common throughout the universe than solid, liquid or gas. From this we know that we have in the past mistakenly supposed that the touchstone for reason was ordinary experience, whereas it ought to have been data. The data come from many levels, not just from one or even two; but there is not a different logic for each level, only different logical relevancies. Imbedded in what is called the experience of the average sensual man are certain generalizations which are held with equal firmness and which later findings seem to contradict. The facts are the facts; and we have in each case to find how the logic applies. Concern with this problem will occur again when we come to consider the interpretation of data.

What is called a datum in science refers to a large area with vague and indeterminate periphery. There are data beyond the reach of observation which are still broadly speaking referred to as facts – data in the broad sense. These are the facts of the past and those of the future. But data in the narrow sense exclude both of these areas. Data in the narrow sense involve repeatable observations. For observations to be repeatable, the observer must be dealing with a class of observations. We can no longer observe the psychological effects of the Napoleontic wars but we can observe the psychological effects of wars. Science more properly is occupied with what can be observed again.

In attempting to get at what science means by data in the narrow sense, we shall have to eliminate the whole of history. History, considered as those things and events which lie in the past, contains facts which are inaccessible to observation and also to experiment. They may be true, and indeed all statements which refer to what actually did happen are true statements; but if we are to understand science in terms of its method they do not lie within the scientific provenance. Thus there can be a science of history only as the study of the laws of development. The attempt to achieve truth and accuracy in historical writing, while stimulated by the spectacle of science and admirable in itself, is in no exact sense a variety of scientific inquiry. The data in the narrow sense to which science refers are not those of the past.

In the narrow sense of data, as classes of occurrences based on observation whose existing is irrefutable, past facts would have to be allowed because *based* on observations in the present though themselves no longer observable. In this class belong the carbon 14 datings of fossils and artifacts, the chemical analysis of ancient documents, and the "supernova" of A.D. 1054 as recorded in the Chinees and Japanese manuscripts of that period. Whatever has happened has happened, and nothing can ever occur to alter the character of past happenings; but ascertaining what it was that happened – knowing it *as* a datum – present enormous difficulties to the observer, who must infer from what is present the irrefutable elements of the past.

There is thus only one context in which the past is relevant to science, and this is in the way of general theories concerned with the temporal order. There are, after all, historical hypothesis, and the developmental sciences are devoted to them. Theories of cosmogony, of evolution, of genetics, and of development in other fields, are certainly concerned with the past. However, they exist and operate as sciences very much in the present. An archaeological fossil, a genetic mutation, the red shift considered as an argument for a single cosomogonical origin – these are evidences or facts in the present which are adduced to cast light on the past. But as data they exist now and not in the past.

The facts of the future are relevant to science in much the same way as those of the past. A future fact is one which scientist cannot observe. The relation of future facts to experiments is something else, and makes facts of the future of greater scientific interest than those of the past. We can control the future to some extent, and from our observations of facts in the present we are able also to predict. However, as we found to be the case with past facts, future facts are the objects of present theory. Present facts, and these alone, are the observable facts. Others can only be connected with the present through theory. The future holds a slight edge over the past in that it has yet to go through the present. Neither past nor future facts are observable, yet there is a vast difference of emphasis with respect to observation when we distinguish between what is not observable because it is *no longer* observable and what is not observable because it is *not yet* observable. The last visit of Halley's Comet, in 1914, will never be observed again, while the next visit of Halley's Comet, in 1986, will be observed by those who prepare for it. We know about both, but the relevancy to present-day observation is not entirely the same in both cases.

(c) *Conceptualized particulars*. The order in which data are collected is not the order to which they belong. Moreover, the very recognition of that which is observed required the use of an appropriate language. Thus all facts are facts in a system. The phenomenology of a domain is a statement – if possible mathematically expressed – of the observables and their physical (chemical, biological, psychological or cultural) interpretations. The relatedness of things insures that there shall be none which are absolutely independent, yet by a datum is meant something intrinsic, to be considered solely on its own ground. In the description of data, then, the observer is committed to the involvement in a language system of an item which he had wished to consider apart.

There are no isolated data which can be accounted for as such. Yet data are individual, and there are no sentences which completely describe individuals. And yet *starting* from the data and being always held *referable* to the data does not mean being *confined* to the data. In a process of abstraction such as the lowering of empiricism called for in the scientific method, material objects and events are treated as the values of variables. Every individual is of course unique to some extent, but it is one of the scientific axioms that there are no absolutely unique individuals, no actual material objects which are unlike all others in every respect. If there are individuals, then there are classes, and once given classes they can be related systematica-ly. Thus the scientist can build his data into systems easier than he could completely describe any datum.

Consider the situation not from the point of view of fact but from that of language, and the same conclusion is developed. For instance, it is true that no sentences can be constructed without the use of generals. Most words, technical words as well as words in common use, are the names of classes, and none in science accurately identify individuals. Moreover, no number of generals (i.e. class-names) can specify an individual. Hence the description of an observed fact becomes exceedingly difficult if not impossible. Mere observation is not enough for scientific purposes; we wish to have a report of what was observed – the individuals. But reports rarely do just that; other than dating and placing (and usually just dating), reports of scientific observations usually indicate generic situations. This has the disadvantage of being non-specific and the advantage of being repeatable: *that* event can never occur again but that *kind* of event can be observed by others desirous of checking the observation.

Thus every datum once described is immediately assigned a place in two systems: the system of the descriptive language, and the system referred to by that language. These ought not to be confused. Modern logicians are concerned that we shall know whether assertions are *in* a system or *about* a system. The former are made from within and employ the object language of the system, while the latter are made from without and employ a metalanguage. It should be remembered that every language *is* a system, and that the grammar of the scientific language is mathematics. Science learns about its subject-matter from the observation of individuals and describes these in the empirical language but soon moves to general statements couched in the mathematical language. We shall note more about this later. Here we have been talking about the formation of the empirical language, dealing with its concepts as these are formed by assigning names to abstract objects. At this stage the efforts of the investigator are directed toward conceptualizing the environment. In the next chapter we shall see how he constructs theories by combining concepts.

Suffice to say that science would not be interested in isolated events even if there were such things. For science is an ongoing process and concerned to find conditions under which events occur, not the events themselves. All experience which takes place through the senses discloses the existence of particulars which are in some ways unique. But sets of them are also in some ways similar, and it is these similarities which suggest the existence of abstract classes. The investigative method of science begins with the sheer awareness present in passive observation and ends its first stage at discovery of conceptualized particulars. A conceptualized particular is a description of an individual. It is a generalization from sense observation for which some term of description has been found. It could be either a familiar term used in a new connection – given empirical support, for instance – or a newly coined term to indicate the same sort of evidence.

Conceptualized particulars have been called concepts, and the way in which the scientist learns about them concept formation. Concepts formed out of the similarities of sets of particulars are themselves abstract. They are no less abstract because he has learned about them by means of his inspection of material particulars; they have, it is always hoped, an *a priori* status in no wise dependent upon his having acquired knowledged of them *a posteriori*. And when they are combined into empirical laws and these laws tested back against experience, it is with the hope this time that by this method he will be able to confirm them in their status as *a priori* principles. Thus data are interesting if they lead to the other parts of some system in which they belong. What is suggestive at one stage of science may not be at another. What is significant in the way of fact depends very much upon the given stage of science at the time. Galileo's experiment at the Leaning Tower of Pisa would be of no significance to a generation which had gone as far with the theory of gravitation as the present one. The discovery of data is only significant when they reach beyond existing theory and suggest further observations, inductions and hypotheses.

Data are included in systems by means of the theories which account for them appropriately. The scientist is ready to abandon a theory when the data collected demand that he does so, but that is only because the existence of facts is easier to support than the confirmation of theories. When generalizations conflict with the phenomena, he "saves the phenomena." For the confirmation of theories is never more than approximate; there are no theories absolutely confirmed. Conceptual inferences from perceptual data are always based on the insufficient evidence, never more than lucky guesses or happy inductions. The scientist endeavors to interweave data and theory in order to achieve a respectable set of generalizations which have experimental support, but he does not ever regard his generalizations as final or as testimony in the face of the phenomena.

Where, then, is the solid ground? The question assumes the existence of absolutes but the answer leaves open only the search for them. The less well instructed among investigators, forgetting the languages in which all formulations imbed data within systems, assume that the data are the dependable items. And so, relatively speaking, they are; and yet they are not absolute. When a statement which purports to refer to facts differs from some fact, he keeps the fact. Data are always true, they are never false. What may be false in the theoretical context in which he may have erroneously supposed that they belonged. One could easily misinterpret a fact, but that would not be the same thing. Errors with respect to fact are those of interpretation, not of the datum itself.

However, although the scientist takes his stand upon data because they are relatively more reliable, he must not allow himself to forget that data are parts of systems and that the systems themselves are not absolute. Negative facts may even be inimical to the systems in which they have been proposed. There are no sacrosanct systems within science; any formulation may have to be abandoned. And so the final outcome is that the scientist is on shifting ground in dealing with variables as though they were no more than a collective name for their values. Elementary pursuits in science are delicate affairs, and as liable to lead him astray as the more advanced problems and inquiries.

Concepts in science are formed on the basis of observations and then fitted together into systems. The observations are made at the level of concrete particulars, of individual material objects in motion in time and space. The concepts are formed and combined at the level of abstract universals, general propositions whose recurrence shows that although they were suggested by concrete particulars they themselves are not concrete particulars nor in any way confined to individual material objects in motion in time and space. Science is the search for those abstract structures which are suggested by material objects; and thus the enterprise of the scientific method is one of weaving the two levels together in a world where uncertainty is always an element and there is no solid ground which can be said to lie altogether beyond errors of observation and of interpretation and therefore beyond all dispute.

Our next task, therefore, is to see how the investigator arrives at his abstract structures, and so we begin with the simplest and watch the construction of classes.

CHAPTER III

THE SEARCH FOR HYPOTHESES: INDUCTION

Science, as we have noted, devotes its first stage to the observation of data, at first passively but then under controlled conditions. It confronts the data with the commonly received notions which are the inheritance of common sense (that traditional philosophy of experience). Then it moves to make the corrections which are forced on it by the new data. This last step involves some degree of theoretical speculation. The rhythm of scientific investigation calls for periods of immersion in fact alternating with periods of withdrawal to consider the observed material. The first stage was that of observations, of samplings, ending with conceptualized particulars. Further samplings locate interpolations between the observed facts to reach smoothed out generalizations. The first half of the second stage consists in collating data and then by their means proposing the generalizations as invariants. It may be named construction. The discovery of hypotheses mark the second half and is named induction. We shall begin by examining construction.

1. *The derivation of classes*

The aim of empirical science is the discovery of logical structures which do not extend beyond the observed facts any more than is necessary in order to account for them. But there are many steps between. The road to such discovery properly begins immediately after observation with probing operations, the consideration of classes more or less tentatively, but then goes through many other steps before arriving at a body of related mathematical laws.

We shall see that (a) *construction* involves (b) *abstraction* from observed properties which makes possible (c) *classification* necessary for (d) *description* and (e) *definition.* This is not necessarily an ordered series of operations but more often a set of alternative ways of approaching a problem. It is a graded series from which selections are made according to the type of procedure chosen.

(a) *Construction.* The stage next after the search for data by means of observation in the empirical dialectic which constitutes the scientific method begins with individual material objects, well below any logical forms, and then relates the individuals to the forms. It consists in the discovery of a class by means of the comparison of individuals, and the naming of their similarity. The process does not end there but goes on to connect the classes. The movement from material objects to classes related to other classes is what is here called construction. We shall see how the process develops in a number of steps, from abstraction and classification through explanation and expression, until it terminates in inductions to hypotheses.

According to Wittgenstein's account in the *Tractatus,* facts are named and the names combined into "atomic" propositions which are pictures of facts. The atomic propositions are next combined into "molecular" propositions and the molecular propositions themselves combined in ways which mirror the larger segment of the natural world. These, as we shall note shortly, are of necessity far less orderly; for in this part of the scientific procedure the investigator is governed to some extent by chance encounters, unexpected discoveries, and surprising perplexities.

The construction proposed by Wittgenstein involve a primitive kind of logical movement which occurs early in the scientific method. To climb from material objects (or facts) to classes of material objects constitutes a basic step. For material objects belong to many classes, and the choice of a significant class is involved. Insignificant classes could be chosen in the same way, and by the same method, by investigators lacking the training which is a prerequisite of the scientifically useful.

Thus construction is a species of technique. It consists in ordinary operations conventionally performed, the journeyman work in the sciences, and is usually not dignified with a pretentious name. Neither is it a process reserved to genius and descriptive only of the largest and most inclusive theories, but a generic term for some simple operations involving processes of thought: reducing data to knowledge, placing new knowledge properly among old knowledge, fitting cases under suitable classes, selecting abstractions to cover data. Each of these steps relies for its guidance upon the process of building from the individual instance to the general formulation. Inductions, as we shall presently see, are more complex and apt to be more important. These are the genius operations, when broadly inclusive concepts are involved. We shall confine ourselves first, however, to the construction of classes, and reserve the discussion of the formulation of inductions to the second half of the chapter.

(b) *Abstraction.* The procedure of scientific method begins with immersion in facts by means of observations. But we have said that it calls afterwards for a withdrawal from observations in order to consider the observed material. By abstraction here is meant the representation of the partial aspect of a material object by means of language. Abstraction makes possible further observations from a different angle. The absence from facts is provided for by means of abstractions, which are, however, always derived from facts and always held answerable to them.

The process of abstraction, then, works upward from concrete facts to the language in which they are expressed. Facts are particulars, the individual instances of classes of facts. The language in which facts are expressed is inherently abstract. Many attempts to escape from the abstractness of language have been attempted yet all have failed. Proper names are an exception. Descriptive language in the effort to account for facts is compelled to use abstract terms, and the current locutions undertaken to return to individuals from abstract terms are highly involved.

The aim of science is always to use the individual instance to discover the general proposition (or equation) which will not be limited to that instance but will instead be true of all such instances. In other words, in order to arrive at knowledge applicable to the concrete world, such knowledge must be general; otherwise it would be time- or space-bound and applicable only to some particular occasion. There is no science which treats merely of individuals, but there is a method of holding principles down to individuals, and that is the scientific method. It is not possible to move from individuals to other individuals without the use of generalizations, as Mill thought could be done. The movement from individual to general is inductive, that from general to individual deductive. The general is a way of getting from individual to individual, and there is no other way; and if we do not as empiricists like the idea that every general extends beyond both the individuals which suggested it and the individuals to which we know it applies, we shall have to settle for it anyway, and hope to extend the testing, for we cannot eliminate the extension.

Such abstractions are independent of the observer and takes place as it were objectively, not merely within the subject making the observations. Scientific abstraction is not psychological abstraction. It is closer to what Peirce called "hypostatic abstraction" and Whitehead "extensive abstraction." By "hypostatic abstraction" Peirce meant that the capacity of a percept to assume propositional form in a particular deduction consists in the truth relation between concrete subject and concrete predicate, the abstraction which makes it possible to consider a fact under the form of a relation.

By "extensive abstraction" Whitehead meant that in a world composed of events related to each other as wholes to parts, classes can be derived from events by analyzing a large event into a convergent series of successively smaller events until such convergence to simplicity leads to an abstractive set (i.e. a class). The role of the investigator is to make the abstractions, but they are made from the data disclosed by his experience, not from within his experience itself. If science is the method of extracting invariants from substances, then the invariants, which resemble logic and mathematics, yet are as much part of the material world as the substances from which they were extracted.

Abstractions may be of two types, and these are often confused. The first type consists in abstracting from material objects. The second type consists in abstracting from the abstractions.

Abstractions of the first type consist in deriving from the material world those elements in it which have a common property. Hence classification, as we shall presently see, is more truly abstraction of the first type, for it moves from material individuals to abstract classes.

Abstractions at this level may be qualitative, relational or quantitative.

In 1823 Herschel suggested that the bright colored lines against a dark ground, which was the result when light from flames tinged with metal passed through a glass prism, might be used as a test for metals, is an example of a qualitative abstraction.

Joule's Law for the transformation of heat into work

$$W = H - h,$$

where W is the amount of work, H the amount of heat from a source, and h the part given up, is an example of a relational abstraction.

Einstein's equation for the equivalence of mass and energy

$$E = mc^2$$

where E is the energy, m the mass, and c the velocity of light, is an example of a quantitative abstraction.

Quantitative abstraction is more typical of scientific investigation. In quantum mechanics, the results of experiments like the photo-electric effect of 1900, which indicated that electrons had the discontinuous structure of particles, and the results of experiments like that of Davisson and Germer in 1927, which indicated that electrons had the continuous structure of waves, were related by Planck's quantum of action, h. Such abstract compounding of classes is apt to be almost altogether quantitative. This is true certainly when more complex theories of physics are stated, such as the special and general theories of relativity.

The higher the level of abstraction achieved by a science the more successful the science, provided only that the correct method of abstracting from concrete fact has been followed. Starting from the abstractions from concrete fact and moving on to abstract from these abstractions, the method of scientific investigation leads inevitably to mathematical formulations. The stages of abstraction run from repeated experiences and observed uniformities at the concrete level, to invariantive uniformities and statistical probabilities as halfway stages, and then to mathematical tautologies as the final stage.

Thus abstraction of the second type is a method of progression, leading upward to the highest mathematical levels. But of course complete mathematical expression is impossible. In science there is always an empirical content, not present in the formulations of pure mathematics. Abstractions at this level serve as background networks into which the investigator continues to fit the data which his new experiences disclose to him. Mathematical usage in empirical science always refers to substance, to matter or energy; it is never completely abstract, as are the formulations of pure mathematics. For no matter how high the abstraction if it is empirical it still depends eventually on concrete facts, just as the top storey of a skyscraper is supported by the foundations on which the whole building rests. An actual reference is still there implicit in the most abstract mathematical formulation so long as the latter has been obtained as the result of a correct following of scientific procedure.

It has been argued that all abstractions is falsification, and this is to some extent true. When we abstract we have regard for just those properties of objects which they have in common and by means of which they are recognized as similar, and we abstract from all other properties which may tend to render the object specific and singular. Thus what we take for our consideration and manipulation was really there in the external world, though it was not all that was there. In any connection in which it may stand for all that was there, it is truly a misrepresentation. The abstractions of any science have to do not only with a selection of this sort but also with a wider selection in the sense that the objects chosen for classification, description and definition belong to some particular area and do not constitute the whole of nature. Inasmuch as every natural object is in the province of some one empirical area, each science consists in an arbitrary abstraction, then abstracts further in considering only the general properties of the objects in that narrow area. The necessary falsification which is always involved in taking the part for the whole is more than compensated by the degree of precision and inclusiveness which is thereby attained.

(c) *Classification*. Scientific method begins with the observation of data.

It moves on to abstraction and the defining of classes of objects. The next stage is the combining of names in propositions, of which the first subdivision is classification.

Classification is the naming of properties, the discovery that items having a certain property form a class. The investigator has, say, a set of material objects, and he finds that the set has a common property; he names the property, and then considers the name as the sign of the class. All further objects discovered which have this defining property he regards thenceforth as members of this class. Thus when a newly discovered object is to be classified, this merely means that he will place it with similar objects having the same defining property. Individuals are grouped into classes because they have similar properties, and the classes are joined in more inclusive classes. Thus genera are groups of species, from individuals (infima species) to wider classes (the genera) theoretically until the widest class is reached (the summum genus). Classification thus climbs upward from the individual to the general. An object may belong to more than one class. Objects have been discovered by means of classification working in the converse direction, that is to say, climbing downward from a network of classes until the particular is reached, and then finding an individual which answers to the description of such a particular.

Classification is possible only because there are similarities between entities. Objects present similarities as well as the differences to ordinary observation. The classification of similarities observable under conditions of ordinary experience occurs in the early stages of science. Thus it is not much above the descriptive level. Science at its elementary descriptive stage classifies objects whose properties are observable to the unaided senses. It does not seek for properties whose presence is ascertainable only with the aid of instruments or which lie at deeper analytical levels. The properties most simply observable are qualities: hardness, color, smell, and so forth.

But other properties of a complex nature, or even functions, can form the basis for scientific classification. These are intuitively selected. Biology for instance is divided into botany and zoology. The plants are divided into four main divisions: *Thallophytes,* simple plants without roots, stems or leaves; *Bryophytes,* liverworts and mosses, with stems and leaves but no roots; *Pteridophytes,* ferns; and *Spermatophytes,* seed plants. As an example of the animals, let us follow down through the subaltern genera until we come to man. We move from the *Phylum* (or division) *Chordata* to the *Subphylum Vertebrata* to the Class *Mammalia* to the Order *Primates* to the Family *Hominidae* to the Genus *Homo* to the Species *Sapiens.*

Such crude tentative classifications are meant to be no more than beginnings. As science probes deeper, its first judgments may be changed and its early classification found to be no longer useful. The Greek classification of the elements: earth, air, fire and water, was still used in the Middle Ages and proved inadequate only when chemistry began to make finer distinctions. The four elements of the ancients were later replaced by one hundred and two; they were originally classified by atomic weight, but since the increase due to the discovery of isotopes they were now classified by atomic numbers.

The world presents us with objects having various degrees of similarities to and differences from each other. This is as true of events as it is objects. Our first task if we are to deal with this world is to look for signs of order. Now classification is the recognition that there exists a certain rudimentary order. If there were no similarities, then there could be no classification. The investigator groups together objects which are similar, and sets apart those which are different. In this way, he begins to recognize order and disorder. He is, in other words, seeking to discover classes rather than to invent them. Scientific classes are those resulting from empirical investigation.

Let us suppose that the truth of the proposition "All swans are white" had been regarded as established and that then some black swans had appeared. Now, the question is, when they were discovered was the proposition "All swans are white" falsified, or did the word "swans" acquire new meaning? It hardly seems satisfactory to say that the assertion "There are black swans" meant only "Some birds exactly swans in all respects except color are black." Whoever is tempted by this explanation, as opposed to the scientist's conviction that he has made a discovery which falsified an old belief, fails to recognize how important is the notion of significant classification for science.

There is an objectivity to the process of classification which cannot be escaped if the process is performed correctly. For the investigator does not set up a class name and so make things alike; they are alike, and that is why he sets up a class name. He is guided in this also by the logical rule that the classes must not overlap. Classification seeks out the natural order, an order which must have already existed among the material objects in nature. But the finding of this order does commit observation to a certain degree of interpretation and involves it with a certain amount of interference. And it may eventually lead to experimentation.

The question of whether classes are natural or artificial, that is to say, whether they are discovered or invented by investigators, cannot be settled. The aim of all scientific inquiry is to discover order in the natural world.

Some formulations that are assumed to be natural prove on further inquiry not to be so; thus classes thought natural may turn out to have been artificial. Certainly, however, if things or events possess the properties we recognize in them, and if we use those properties as the names of qualifying characteristics of classes, then the classes may be said to be natural. A difficulty arises in assuming that there can be only one natural class. This is not the case. The same material object may be a member of a number of natural classes. The same man may be as much a taxpayer as a father, as much a wage-earner as a voting citizen or a property-holder.

The question of whether classes are natural or conventional must be crucial to the larger question of whether the scientific method is one of discovery or of invention. A conventionalistic philosophy of science would assume that classification is a human and hence arbitrary decision resting on subjective grounds within the investigator. The view assumed in the present work is that this is not the case. There are many reasons for preferring the objective position and so for arguing that objects do belong to the classes to which investigators assign them. Two may be mentioned briefly here. The investigator is not concerned with his own psychological processes but instead with what those processes can reveal to him concerning conditions in the external world, otherwise he would not need elaborate physical instruments and mathematical operations but only introspective moments. Again, the self-corrective nature of the scientific method comes to our aid here, for it is in terms of what the investigator holds to be the true conditions that he corrects his findings and not in terms of his own arbitrary changes in preference.

What the scientist is seeking, of course, is significant classifications. In all scientific endeavor, the best choice is usually that which leads to further investigation, that which opens up lines of inquiry. Some classes, then, have a higher scientific utility than others, and those which have seem to be the most natural. The first business of science is the classification of *provocative* facts and not just merely facts. Later on we shall see this same point developed in terms of crucial data.

The truth is that the selection of appropriate classifications is a matter of the same kind of scientific intuition that determines other scientific advances. We do not gather all the facts before we classify them, and our selection of facts is determined at some stage by the classifications themselves. Thus it would be untrue to say in any absolute way that in science facts precede theory or theories precede facts, although to be sure when facts and theories conflict the usual tendency is to save the facts. For it frequently happens, especially in the more developed sciences, that where facts seem to conflict with theory it is often possible to show that on another interpretation the conflict is an appearance merely and not genuine at all.

The discovery of both fact and theory goes together in an interrelated process. A classification is good if it suggests further inquiry and bad if it does not. A good classification features important properties, discloses significant relationships and introduces new principles. A classification is bad if it imposes a scheme upon the facts which ill fit them. The classification of chemical elements by their atomic weights was good because it enabled Mendeléef to see the periodicity between them. A classification which does not suggest further lines of investigation is soon dropped. Kretschmer's classification of physical types of persons: the *athletic* (strong skeleton, broad chest, powerful musculature), the *aesthenic* (tall and thin), the *pyknic* (short and broad), the *dysplastic* (an "intermediate" group), is taken not too seriously any more, while his correlation of these types with varieties of psychoses is open to much doubt. A classification is not intended to be an end in itself, no more than any other development in a science. Classification enables us to record characteristics in order to perceive relationships, and so to take the first step toward the discovery of important principles and laws.

The theory of classification called taxonomy is not thus far developed. Classification, as we have seen, begins with field study, where it is guided by ordinary careful observation conducted at the level of enlightened but unaided sense experience. It graduates to the laboratory, where it receives the aid of instruments, and is thus prepared to aid in penetrating to deeper levels of analysis. The early classification of all astronomical objects into planets and stars has become outmoded as a result of observations through instruments. Subatomic particles could never have been classified at all without recent method of exploration and controlled experiments.

The classification of objects must not stop but must move on into higher regions of abstraction and eventually into mathematics. Classification is halfway between the objects observed and the abstractions of mathematics. It is in many cases a necessary step, but if investigation stops there an unfortunate step. In zoology, the process of classification has been complicated and comprehensive, but tends to inhibit further inquiry. The scientist needs to move from the discovery of classes to inductive generalizations, and classes which do not lend themselves to higher abstractions in this way are of no great assistance. But this must not be taken to mean that the more elementary tasks of classification are completed. Although taxonomy has been viewed as a primitive stage of science from time to time, its battles are not all won. More than half a million species of the class *Insecta* have been described, and estimates of the total number run as high as ten million.

(d) *Description*. By "description" here is meant the setting forth of the properties that characterize a particular class. The more prominent features are noticed first and they are listed by the method of simple enumeration as matters of brute fact. The investigator is here endeavoring to describe – at the level of ordinary experience at least – the qualities and relations disclosed to his senses and interpreted as attached to the material object. Boric acid is a white crystalline solid. Lignin is an aromatic compound found in trees. The victims of hebephrenic schizophrenia exhibit behavior which is both silly and untidy. The approach to the complete identification of data tends to become gradually more precise as scientific progress is made. It can happen that a description will be in terms of mechanisms or of efficient causes. In all events, something needs to be included of the techniques by which the descriptions are obtained. That ruminants, such as deer, sheep and goats, have no upper incisor teeth might be called description available at the level of ordinary experience, but that hautorium is a parasitic fungus of the living cell is obviously not so. One measure of how precise is the description often appears from the degree of system indicated. A description which inferentially suggests that the object described is a loosely integrated member of a collective is more apt to be the crude level of data of ordinary experience than one which suggests a tightly integrated element in a system.

Although absolute description is an impossibility, when provocative facts are observed they have to be described as accurately as possible. If the fact belongs to a class which is perhaps already known in another connection, then the recognition of its name would suffice. The crystallization of the tobacco mosaic virus by Stanley put the virus in the class of crystals, which was a class already well established. What was new and provocative was that the entity recognized as a virus could under certain conditions also justify classification as a crystal.

Description may be of entities, processes or relationships. The kinds of objects described are too numerous to list exhaustively, but a few examples may be offered. Typical of entities described are discrete objects: planets, plants, cells. Typical of processes are: forces, fields, interactions. And typical of relationships are: functions, causes, correlations. There are other types of objects calling for description, such as domains, for instance, and formulas, and even investigative areas. Any of these may require descriptions made up from observations taken directly or indirectly; descriptions made after indirect observations are usually of objects whose existence is inferred from certain events interpreted as results.

For entities and processes never before observed, descriptions have to be provided which are univocal. For this purpose it must be in terms of properties or in terms of the operations under which the entities or processes in question can be observed. An entity or process observed is described in terms of its class, not in terms of itself considered as unique. And this is so for two reasons.

In the first place, science is not concerned with unique occurrences, as we have noted, but with repeatable occurrences. Entities and processes to be of scientific interest would have to be repeated, and a repetition would not be the identical entity or process first observed but instead another member of the same class. If an astronomer, using a sensitive thermocouple with a small circular receiver in the focal plane of a large reflecting telescope, succeeds in getting significant measurements of the radiation of some particular star, his work will be repeated by others in order to check the reliability of his findings, on the assumption that in stellar radiometry the same results should be obtained when the same readings are taken under similar conditions.

In the second place, it is impossible to describe in language an unique entity or process. Language is perforce general, whereas occurrences are unique; and no way has been found in language to uniquely specify an individual other than by naming. We can always assign a name to an individual or we can point to it; and these are the means employed to circumvent the failure of language to specify the singularity of the individual. Naming is arbitrary, and ostensive definition employs non-linguistic gestures.

Description, it should be noted, is by no means to be confused with enumeration. Since the aim of inquiry is to describe essential entities and processes, typical instances are apt to be more favorable than innumerable instances. It is indeed quite possible for too many facts to hamper the construction of explanatory theoretical schemes, as often happens in psychology and sociology. The movement from individuals to classes is an abstractive procedure, not the result of a summation. The observer is satisfied when he has described a fair sample of typical instances and does not need to repeat the observations or the descriptions once he has inductively satisfied himself that they can be repeated.

Description, it should be remembered, is an approach to definition. Long before the investigator is ready for the kind of precise understanding requisite for definition, he has made the acquaintance of a new object and needs to describe it. The description will help him in the eventual task of definition but is not one.

(e) *Definition*. The last task in the construction of classes consists in definition. A definition is a more precise kind of description, one established formally and employed conventionally. However, we must hasten to introduce a caution at this point. We are discussing the elementary stages of experimental science and not those of mathematics. Hence definition will mean something different here from what it does in mathematics. In its early stages experimental science is content with a working definition, whereas mathematics begins with definitions which are logically precise. Experimental science can operate with ostensive definitions, that is, by pointing to the object in question while uttering its name. It is possible to operate on laboratory mice without feeling the need for a precise definition of the species. Euclid, however, did not find that this was true in the case of a triangle in plane geometry.

Definition, then, in the early stages of science involves assigning to a word a meaningful association by one method or another. Provocative cases of phenomena are compared for their similarities and tagged in such a way as to facilitate the recognition of further instances not yet encountered. A definition specifies a class of objects. The purpose of a definition is acquaintance with the defined object as a member of the class, and this holds equally for those not encountered and for those encountered. Names are the means by which language reaches down in an effort to ground itself on sense certainty, and it corresponds in this way to actions with objects where both actions and objects are individuals. This is the first step; a second step consists in the qualification of classes (what Aristotle called secondary substance).

Often an object is named after the man who discovered it. The phenomenon observed when radiating atoms are subjected to a powerful magnetic field was discovered in 1896 by Zeeman and is now called the "Zeeman effect." Disease parasites capable of multiplying only in a host cell have been discovered halfway in type between the bacteria and the viruses, and named rickettsiae, *Rickettsiae prowazecki*, after Ricketts and Prowazek, the two investigators who first described them. If the word is associated with a meaning rather than with an object, then the definition may be one arrived at intuitively. In either case it is simply a question of name-giving. In Dreyer's *New General Catalogue* of stars, maintained by the astronomers, newly discovered items are given numbers to identify them. Subatomic particles are given names which may suggest some one of their relations: "meson" for a particle with a mass halfway between that of an electron and that of a proton, and "neutron" for a particle with no electric charge. But these are the names of classes of mesons and classes of neutrons, not of individuals. Individual mesons and neutrons receive no names. It is significant that no proper names are assigned to individual material objects except at the level of ordinary experience. No proper names are assigned to them at either the microcosmic or the macrocosmic levels. No cells as such are given their own names. The only exception, perhaps, are the stars and the galaxies. The visible stars have names dating back to the period when the only astronomy was that of the mesocosm. All stellar bodies are given numbers now for purposes of easy location in order to continue the study of their properties. New numbering systems tend to stress membership in a class as well as to identify individuals.

Names place a new item of knowledge in the context of knowledge as it already exists. A new item may have to receive a name which shows that while it is new it is also recognized as a species of an old genus. A new kind of fish would receive not an arbitrary name but one which in itself conveys a certain amount of information. Naming is tagging by reference to one or more characteristics. The investigator never knows what it is that he has named. The object in science is named like a child: we never know what it will develop into or what we will find out about it, but meanwhile we have a name for it.

The difficulties involved in names are many. Names are labels attached to individual objects or events in space-time: matter or energy. This is often difficult. How can we distinguish two pennies by their names, when the only difference between them so far as observation can aid us is their difference in location? The indefinite pronouns come to our aid at this point, and we speak of *this* penny. What we are reduced to by this dodge is only a way of pointing. But such ostensive definitions present their own problems. For instance, we must have the penny present if we are to refer to it in this way. In the last analysis, we are unable to specify an individual by means of a name to those who are not in some fashion party to the specification. In ostensive definition, we point to the object and think about it or utter a name, and we consider the name to be the name of that object. But in empirical science, this is never true. For when we point to the object and utter a name, we mean the name to be the name of *such* objects, or objects similar to the one to which we are pointing as well as that object itself. We use the object as a member of a class (assuming that for every object there is a class), on the scientific assumption that no class without an object is to be recognized, and by means of it (i.e. the object to which we are pointing) name the class. We do not name that object by itself. In the scientific procedure, care is exercised to insure that all words or signs employed shall be names of classes of material objects or of relations of classes of material objects; and nothing else is allowed to have scientific validity.

The simplest kind of naming is the naming of observed facts. The observer assigns a name to the reports of his experiences on the assumption that these are independent of his experiences and need to be tagged for ready recognition. The facts named may be entities, processos or relationships, and the names may be fixed by means of description or more precisely by means of definition.

Early definition is a matter of expediency rather than of exactness. If definition at this level is mainly a matter of name-giving, and if naming places an item in a system, then the procedure must be as tentative as possible, since we are not prepared with the system in which we can place the object. In the early stages of naming we are merely making a start at discrimination and not aiming at anything final. We shall be content with a rough-and-ready means of identification. The system itself will be constructed slowly and then we shall be in a position more specifically and distinctly to clarify our definition so that it can approach something nearer to the ideal which is set for definition by logic, which is that the *definiendum* shall be precisely encompassed by the *definiens* and everything else just as exactly ruled out.

A definition in the rudimentary scientific sense here intended is that association of a sentence with a name, an assignment of a predicate to a subject in some manner. The most familiar form of definition deals with the class directly. It consists in placing an object in the class of objects to which it belongs and then in distinguishing it from other objects of the same class, Aristotle's method of definition by *Genus* and *differentia*. A *genus*, says Aristotle, is what is predicated of a number of things exhibiting differences in kind. The *differentia* marks off one kind of thing from other kinds within the same genus. For Aristotle a definition signified a thing's essence. Substance, for example, he defined as "that which is individual," with the genus, "kind of thing" understood.

Modern scientific definition has modified the Aristotelian variety to the extent of employing function as difference. It consists in substituting primary function for term. For instance, "Catalase is an enzyme which decomposes hydrogen peroxide." Operational definitions seek to perform the functional task but often fall short into mere description; nevertheless their employment as definitions in use is to some extent justified. For the investigator uses the definition in order to recognize the defined object when it actually confronts him in the course of his operations. The limitation of operationalism arises only when the assumption is made that the way in which an entity or process occurs is a sufficient explanation of what it is.

It should be clear that in science the intent is to name not our ideas of things but the things themselves. In certain cases of error, it is possible that the two become confused. Yet it is certainly true that the aim is to define not our private impression but objective general classes and to do so with the help of the properties of individuals. The investigator is occupied with the study of the external world of nature and not with his private reactions to it.

It is paradoxical that while we wish to name the particular thing, names are nearly always names for generals. We may resolve this by saying that names have direct and indirect references. Names always refer directly to generals and indirectly to individuals. The problem of selecting names for observed facts is subject to the stricture that we are not naming an unique observed fact but instead the class of which it is a member. There is of course always a member, for it is the individual considered in its relation as a member of a class which has suggested the necessity for naming the class. The null class in empirical science is represented by the absence of an expected individual.

The direct reference of a name is to a class of objects. "Electron" is not the name of an object but of a class any member of which will serve as the referendum. The names for classes of concrete objects have a more concrete association than the names for classes of abstract objects, but they are equally names and hence equally abstract. What is true of objects is equaly true of processes. "Electrolysis" is a process which can at any time and place be rendered concrete given the necessary materials and techniques. But the word is the name of a class and not of a single individual or unique event.

The indirect reference of a name is the individual entity or process to which it corresponds. If an investigator were to say, "Procure a squid for experimental purposes," it would be not one particular squid to which he was referring but any member of the class, squid. And the point here is that any individual squid in good condition would serve as well as any other, and this would surely not be true had he been referring to one specific squid. Names locate objects as members of classes. The triangle of reference runs from the name through the class to the individual – always to one individual but to no particular one. Thus science from the outset is an abstract affair, incurably general; and it is concerned with individuals only to the extent to which they suggest generals.

Some facts are observed only indirectly through their directly observed effects. These are named heuristically, and are considered to be hypothetical. It is often necessary to account for some effect which seems to be the result of an entity or process as yet not isolated. A naming of the entity or process follows, even though it has not itself been observed. Many of the entities of psychoanalysis fall under this heading, such as the "ego," the "superego" or the "id"; and so do many of the entities and processes of quantum mechanics. No one has yet observed a photon, an electron, a magnetic field of a Schrödinger wave-function. Yet these are viable units because of their explanatory values.

It should be remembered that hypothetical entities are not mere matters of language; they were named in the first place on the supposition of the existence of materially existing particles or events. Either there is something corresponding to the hypothetical entity or there is not. It is presumed that more extensive investigation, more observation and experiment, will decide. Some hypothetical entities are later verified on the basis of experimental evidence and some are falsified. Dirac predicted the existence of the positron, and experimental work bore him out; but on the other hand the luminiferous ether posited by Maxwell to support the undulations of the electromagnetic waves has been rejected in recent decades, especially since the negative result of the Michelson-Morley experiment to determine the ether drift. The employment of hypothetical entities and processes has been increased in quantum mechanics, where the observables are reduced to a minimum because of interference from the instruments of observation.

No distinction is drawn here between material objects and relations, for the reason that both are observable and both need to be named. The yellow lines in the sun's spectrum lie between the green and the orange lines, and this "betweenness" is just as observable as the green and the orange. If it is true that in zoology symbiosis is the name for the association with mutual benefit of two dissimilar species, say the association of termites and their intestinal protozoa, then this relation, which rests on the advantage of food gained by both species, can be observed, and so symbiosis is the name given to the generalization of this series of observations. The investigator observes the phenomena in the external world, and he gives names to items if they persist or recur, and this can be as true of relations as of material objects, or, put another way, as true of classes as individuals.

The facts may be entities or processes; that is to say, the names may be names for something static or active that we can isolate, say, an organism and the process of electrical conduction. In either case, we are endeavoring to attach a tag to something distinguishable so that we may be able to recognize it again. Here we have a concrete object for which we need a name. Even if we wish to manipulate the individual, we have to be able to call its name. And if we wish to manipulate it *in absentia* we shall have to have a name for the type, an abstract name. The thing itself may be recognizably static, an entity; or it may be an ongoing process which we have discriminated against its background.

The world itself is a flux in which everything is mixed with everything else through immersion in time and space. When we describe entities or processes we draw lines around them. Thus a process is a segment of existence, qualitatively demarked from the remainder of existence at both of its ends. An entity is a distinguishable element of the process, regarded as, for that purpose, quasi-permanent. Hence an entity is static, a process dynamic. There is of course always some unavoidable falsification involved in every degree of isolation necessary for description.

Another kind of naming found in experimental science is the naming of empirical ideals. These are essential in the construction of systems and their operation in many sciences. The "ideal black body" of thermo-dynamics, the "frictionless surface" of physical mechanics, the "absolute zero" of temperature, "economic man", the "straight line," the "dimensionless point," all are examples. Hypotheses enter into the selection of the ideal objects to be named. There simply is no such thing as an incompressible non-viscous fluid. It is hypothetical that the objects named are the significant ones (they need not be). And, similarly, hypothetical procedures occur when propositions are constructed by combining names in the scientific language, for they need not be related in significant ways.

It is important to remember that the problem of definition does not present itself only in connection with newly discovered entities and processes but also in connection with problems. The problems themselves may be newly discovered ones, in which case a proper definition of what is involved can be conducive to the correct search for a solution; or it may be a restatement of an old problem, in which case there may be fresh progress toward its solution. The more precise re-defining of problems may be informative as well as suggestive, for the investigator is restating his concepts to get them ready for classification, usually by repeating observations to discover additional properties in order to enrich descriptions. We know more about the organism when we turn from a definition of life in terms of its subsidiary functions, namely, growth, self-repair and reproduction, to a more fundamental definition in terms of its primary function, self-determination.

2. *The formulation of inductions*

There is no compelling or necessary movement of stages from observation to induction. The construction of classes must occur in an atmosphere of freedom, for inductions are reached not by carefully accumulating the data of experience but by widely indulging in imaginative conjectures, and these require that speculation be kept open. Here we need to introduce the idea of random search, for a great deal of what happens at this stage bears this description. Exposure to the material often produces a happy induction, just as often perhaps as the discovery of a provocative fact. Inductions occur plentifully, an important induction is something else again. Most such conjectures are destined to be abandoned after they have been tested experimentally; but it does not matter, for it is from among their number that the theories and laws of nature are to be found.

In Wittgenstein's account of the scientific procedure, the method of construction alone is sufficient to account for the abstractions of a developed empirical science, and either there is no induction at all or concerning it one should remain silent. Here, however, while his description is assumed to be correct as a description of the early stages of science, a secondary stage of induction interposes itself between construction and the adoption of hypotheses. For in the first place experiments would not be needed to test construction carefully executed, and in the second place scientific inductions are not made from the data of observation but rather from concepts.

Inductions henceforth will mean the choosing of hypotheses, and will be referred to as hypothetical inductions. In future chapters we shall be dealing with other kinds of inductions, such as statistical inductions of empirical probability in Chapter V, with the discovery of axioms, the discovery of proofs of theorems in Chapter VI, with tendencies, correlations, and dispositions in Chapter VII.

It is one theme of this book, then, that inductions do not take off directly from raw observation but rather from ground which has been carefully prepared by working over the data of observation. The instance or instances which furnish the particulars to induction are refinements of the data by various kinds of construction: conceptualized particulars. By the time the advanced stages of construction have been reached, there is a qualitative change in procedure, and a fresh start is made, which, for want of a better name, has been termed induction. This consists in a strategy, an impatient construction, lying above elementary construction and extending it. Inductions can be considered dodges taken as the only way out of a difficulty which an abstractive set has precipitated, the leap away out of an impasse. Instead of the discovery of a class by means of the comparison of cases and the naming of their similarity, there is the relation between classes requiring the classes themselves as second-order cases and a comparison and naming done in the same way. Roughly speaking, construction prepares the abstract classes while induction relates them.

It may be helpful to compare construction, induction and deduction as they occur in the scientific method. Induction follows and builds upon construction and prepares the way for deduction. Construction and induction are similar in that both are movements upward in generality. Construction moves upward from the individual facts disclosed by sense experience, while induction moves upward from conceptualized particulars. Construction is a movement from the member to the class; it proceeds through a hierarchy of memberships. Classes of singulars become singulars to the classes above them of which they are members. Induction is a movement from the property exemplified in the member to the genus of which that property is a species, a step based on the relation of being included. Deduction is the relation of inclusion downward from the general to the less general. Deduction requires only the abstract objects of logic and mathematics, although scientific deduction, as we shall later note, is not pure deduction but has a content which consists in the fact that abstract objects refer to material objects by means of deductive form. Construction makes induction possible, and induction is required for deduction. Some deductions (including those in which scientists are interested) can lead to consequences purportedly testable against material objects initially disclosed to sense experience. Thus the investigation which employs the scientific method has come full circle: from the original suggestions in the material world back to the confirmation of the material world.

As for induction itself, a confusion of three areas exists and these call for careful distinction. There is the psychology of induction, the logical process of induction and the philosophical justification of induction. It is induction we need to consider here, not the psychology of induction, which requires separate treatment. In the abstract presentation of the scientific method, the effort is made to render induction less intuitive and psychological, and so to make it available to more investigators. However, even if it is possible to tell what happens in an induction it may not be possible to tell how to have it happen in any important way.

The distinction between the process of induction and the justification of induction is clear enough, for the former is an act whereas the latter is an argument. There is no canon to which anyone can appeal in the attempt to distinguish valid inductions from invalid ones. The presuppositions of induction, postulated from Hume and Mill onward, may be there but they are no assistance in the scientific method.

We are concerned here to discuss chiefly what induction, in the logical sense, is, and then briefly to undertake a justification before introducing its reason for being, which is the induction to hypotheses. But first it will be necessary to approach it through (a) *inquisition,* (b) *explanation* and (c) *expression,* (d) *logical induction,* before arriving at (e) *the justification of induction,* and (f) *induction by hypotheses.*

(a) *Inquisition.* By inquisition is meant the setting of a problem. The construction of classes gives rise to problems, and it is the attempt to solve the problems that inductions occur. Such inductions are efforts to grapple at first hand with the aberrant nature of intractable material. The solution of early difficulties often gives rise to new problems, and the examination of seemingly inconsistent data often does also. There are at this as at every stage always doubts, questions, uncertainties. No scientific investigation was ever undertaken except in terms of a problem. It may be a familiar problem or one only suspected to exist. In the former case a new approach is derived which may make it possible to attack the problem in a fresh way. In the latter case it may be simply that some familiar technique has been found suitable. Exploratory investigations then ensue.

In order to see in more detail just how this works in science, let us look at the way in which conceptualized particulars, as defined in Chapter II, 3, (c), arise in the course of investigation. The transformation of data into conceptualized particulars can be accomplished in one of the two ways.

(i) Resemblances may be perceived between two or more individuals. The nucleus of a cell was observed to divide as early as 1873 and was named mitosis by Fleming in 1879. The need for the name is evidence that the process was recognized to be common to many instances and would be encountered often.

(ii) An individual may be suspected to be typical. In psychology the existence of the eidetic image, which is an unusually clear image perceived as though it were external (common chiefly to children) was discovered and named by Urbantschitsch in 1907, and confirmed by later investigators.

How do mitosis and eidetic images fit into the biological and psychological systems respectively in which they ought to belong? At this point it is important to notice that the conceptualization of particulars is never more than partial, and this for two reasons: first, the knowledge of a particular is always limited and incomplete; and, secondly, it is to this extent a distortion, for in the kind of operational synedoche which the employment of conceptualized particulars involves, the part is taken for the whole. It is, then, the unconceptualized part of conceptualized particulars that raises the difficulties in the practice of the scientific method. We know that there is division of the cell nucleus, and we know a great deal more about cells, we know, for instance, that each chromosome duplicates before the process of mitosis begins, but there is still more – vastly more – that remains unknown; and the same can be said of eidetic images: we know that it lies within the capacity of certain individuals to produce images which are lively and which have a pseudo external character, but we do not know the cause of the mechanism which produces them or what their possession means in terms of pathology. Whenever conceptualized particulars are made up of unfamiliar data, then they constitute a problem by themselves. They need to be assigned a place in knowledge, which means somehow built into a wide system and if possible explained. They may even raise further problems requiring inductions to new theories.

In addition to unfamiliar data, there is also the unknown. Between familiar data may lie unknown entities, processes or relationships. These may have to be posited and then introduced into the investigation to serve an heuristic purpose. Any datum, in short, which is not accounted for by the available theories constitutes a scientific problem. Inquisition formally introduces a line of investigation; for it is only when a problem is recognized as a problem that the scientist goes about determining the best approach to its solution. He needs to state the problem as clearly as he can in order to consider what might be the alternatives; a best approach can be selected only from a number of proposed approaches. Einstein's recognition that Newtonian mechanics did not account for certain facts, such as the Fitzgerald contraction and the Lorenz transformations, led him to the problem which he solved by positing new physical theories.

In developed sciences, systems of theories and laws already exist. These have been moved ahead on a broad front, but leaving many gaps in their wake. Progress is seldom uniform nor made from a well-consolidated base. Thus the scientific investigator may find his problems at any level of abstraction; he may find them among simple facts or at higher abstractive levels.

But at whatever level he finds them it is important to remember that they are not primarily language problems but problems about the subject-matter discussed in language. Scientific problems are treated in the scientific language but they are scientific problems primarily and language problems only secondarily. Moreover, there are limits to the advantages to be obtained from clarity and precision. The presence of a theory in the background may make itself felt as indefiniteness of formulation. There is a suggestiveness to vague generalities not to be found in exact formulations, and while vagueness as such is hardly an acceptable ideal, it may be the most strategic way to the discovery of a proper solution.

(b) *Explanation*. By "explanation" here is meant the proposal of a theory from which a datum follows. Classification, then, is the preparation for explanation. In classification, we have discovered, the principal task is to place the object where it belongs logically. In explanation, we wish to place it where it belongs empirically, to account for it as a phenomenon, and this means to determine how it functions. We are at this stage beginning to deal with the combining of names in propositions. It would not be too broad to say that the intention of science generally is the explanation of nature. The uncovering of the working of the natural world is the primary motive of science.

At the level of explanation we remain as factual as possible and do not need to rise to any high abstractive considerations. Construction, abstraction, classification and description are all called out, but in a fashion which does not admit of any elaborate complications. An explanatory proposition in science has two important functions: it describes the facts and assigns them to some abstract group; and these functions are combined, so that we might say that explanation locates the facts which it describes in some system of relations.

Explanation is the replacement of a more or less inexact description by a scientific one. The description must be similar to that which it is to explain, so that the former can replace the latter in every case. It must be sufficiently exact to fit into a scientific scheme. It must be fruitful, which is to say, useful in formulating general propositions. It should if possible show how to produce that which it explains. How is it made, found, devised, contrived, discovered? Explanation may, more specifically, consist in the argument from effects to causes. This often means going from one level to another below it (analysis). It can be seen from this that explanations usually involve a framework of established or temporarily established entities and processes. The understanding of exchange forces in nuclear physics calls upon the background of the existing knowledge of electromagnetic energy at the atomic level. Thus progress in the solving of a problem will be made possible by the employment of a broader background.

We are not at this stage ready for any elaborate instrumental observation or any complex mathematical formulation. Matters remain more or less at the level of naive common sense. This simplicity serves the purpose of clearing the decks of any extraneous considerations which may otherwise have intervened to hamper investigation. We are trying to perceive what is before us, to describe and account for that which we perceive. Burdened as we are with the accumulation of cultural beliefs, it is not so simple a matter to straighten out the accounts and to admit to our standards of relevance only what is available through the channels of perception and the consideration of only what is actually perceived. For the tendency to perceive what we think we ought to perceive, what, in short, we believe to be there, is a strong one and not easily overcome. Scientific explanation, remember, must account for what is observed, and the purity of this requirement must be retained in some way or other, even though, as we have noted earlier, the category of observables be stretched to include hypothetical facts and empirical ideals.

Perhaps the classic example of explanation is the formulation of the theory of gravitation by Newton in 1666 to explain the phenomenon of falling bodies. Another example is that of vaccination. Jenner had introduced the practice in the case of smallpox in 1778, but it was two years later before Pasteur succeeded in accounting for its success by means of the theory of immunization. Röntgen discovered the x-ray in 1895, but the explanation in terms of radioactivity was the work of Becquerel and the Curies a year later and somewhat later still of Rutherford.

Simplicity in the matter of explanation is a matter of holding the language as closely as possible to the names for concrete objects, or for classes of such objects, together with the barest of the direct relations between them. For after all both complex accounts and simple accounts are conducted in terms of language of some sort, and there is a sense in which all language is abstract. What we mean by a simple explanation, then, is one in which the language refers to objects and their relations and not to other languages, even though, in the latter case, the language referred to may have the same reference to objects.

Such an ideal of simplicity of language corresponding to simplicity of explanation is of course an ideal which can only be approximated and rarely if ever completely attained. We have falsified the situation in talking about naming as a separate and isolated affair in the first place, and to simplify explanation in terms of it would be merely to aggravate the offense. The scientific language may be built up by naming objects and then by combining the names; but on the other hand, all words are by the very nature of language syncategorematic, and meaning does not emerge slowly but all at once, when the names are employed in conjunction with other words.

Once again it is necessary to remind ourselves, however, that the scientific language is not a metalanguage but an object language, and that it refers to material objects in space and time. To talk about the scientific language facilitates discussion, yet accuracy rather than facility is the aim; and it is truth that is sought rather than easy results.

Explanation always involves general propositions in the account of particular facts. We never want to know merely what caused this phenomenon without going on to an explanation in terms of such phenomena. Mill's argument that the explanation of a particular fact can be in terms of other particular facts without passing through general propositions depends upon the assumption that general propositions themselves are only aggregates of particulars. Actually, general propositions are universal statements necessarily extending beyond the instances with which the observer may be acquainted. He does not say merely "These men are mortal" after viewing the dead bodies of Tom, Dick and Harry; he says instead "All men are mortal." Thus the process of explanation depends upon the further support of other instances. In this sense it is a kind of preparation for induction.

(c) *Expression.* We take up the approach to induction with the analysis of the terms of expression appropriate for scientific inquiry.

By "expression" here is meant the incorporating of empirical classes in the appropriate language. The classes themselves as we have seen are derived from the disclosures of sense experience; the logical connectives necessary to incorporate them in a language are peculiar to the language.

The problem, then, is to express what has been observed and abstracted. Language systems are the only means we have in our possession for dealing with facts. The encounter between language and facts puts a strain on language, at least for scientific purposes. The natural languages were developed in response to the need to describe ordinary experiences, a colloquial language for the expression of facts at the level of common sense. But we have seen in a previous chapter that this is where science takes its start, not where it stops. The facts disclosed by experience beyond the level of common sense could be expressed in the ordinary languages but it would not be convenient or economical to do so. The convenience of manipulation requires the kind of shorthand that is supplied by the special language of science, which is always an empirical language and different for every science. Each distinct science has its own vocabulary, and interwoven with each is the mathematical language. There comes a point in the advanced development of a science where the two languages become indistinguishable, and the empirical data can no longer be expressed except mathematically, as in the case now with mathematical physics for instance.

There are still other reasons for holding the natural languages inadequate for scientific expression. The precision demanded in science requires that colloquial language shall be exchanged for mathematical language. Again, colloquial language with its connotations as well as denotations is multidimensional, whereas what is required for scientific expression is one-dimensional. The scientific investigator wishes to translate into language the facts disclosed by sense experience aided by action and thought in a way which shall be precise but carry no overtones. The description of direct observation yields to mathematical expression only when it is sufficiently abstract.

The use of the mathematical language prematurely may be fatal to inquiry, and it is premature when the concepts have been left concrete. The calculus of pleasure and pain of Edgeworth and Jevons (1871) was a failure not because of the calculus, which was employed correctly, but because pleasure and pain are relatively concrete terms, imperfectly abstracted, opposites and not contradictories. Observations which are indirect have a better chance of meeting the requirements of abstractness simply because they are already somewhat abstract. Pointer-readings are abstract to a certain extent by their very nature: the investigator is observing the effects of events on a recording machine and not the events themselves.

The scientific language, then, is finally a skillful composition of the empirical language and the mathematical language, with the empirical language serving as the object language and with mathematics as the metalanguage. Empirical languages are natural languages, not artifical ones. The difference between a natural language and an artificial language is that the natural language is constructed as needed, and its syntax as well as its vocabulary has developed out of the response to actual requirements; whereas the artificial language is a deliberate product of applied logic. Now the empirical language, the language of conceptualized particulars, is for all its technicalities a development of a natural language; while the mathematical language is an artificial one. The problem of empirical science is that the problems, the operations and findings though empirical and so amenable to expression in a natural language need to be formalized in the terms of some mathematical language which has been selected as peculiarly suited to the purpose. Experimentation is an activity at the level of ordinary experience but because it probes into levels of analysis beyond the reach of ordinary experience calls for expression in some more technical and hence more precise language. The terms of description in any language, however scientific, are sure in the end to prove inadequate (however useful in the meanwhile). If the individual character of fact and the universal nature of language are to be accepted, then the formalizations of science, which are always expressed in some language, can never rest solidly on fact but must always ride suspended somewhat insecurely above it, for there is no safe anchorage on a shifting bottom.

Care must be taken, however, not to confuse the scientific language which results from expressing scientific experience in the mathematical language with the artificial languages constructed for the philosophy of science by the positivists, by Carnap for instance. It is an error common to those who work in the artificial languages to confuse the kind of definition which is employed in mathematics with the definition of empirical entities discovered in the laboratory, a confusion occasioned by the concrete interpretation of the former and the highly abstract character of the latter. The artificial languages have the limitations that they are no more mathematical than logic and no more empirical than mathematics, and so fail in the two requirements which any scientific language would have to satisfy. The semantics of artificial languages proves to be logical exercises in the study of language, not inquiries into the methodology of experimental science.

The logical positivists would restrict inquiry to the exigencies of an exact language, and thus would decide in advance the limits of inquiry. But scientific inquiry deals with the empirical world which is full of novelties and for which expressions are always partially inadequate; so that the language of science, like the less exact colloquial languages, must be left free and flexible in order to cope with the data of experience and the generalizations provided for them. Given the abundance and complexity of the contents of the world with which science has to deal, there is no formalized language of science capable of anticipating all possible transactions.

The complete formalization of a sequence of concrete activities such as the scientific method is theoretically possible but cannot be accomplished with the logical equipment presently available. Unless the logical tools are adequate the results of analysis are impoverishing at best and otherwise misleading or even distorting. To do justice to the scientific method it would be necessary to understand the nervous system sufficiently to formalize the observation stage, to understand the cortex sufficiently to formalize the intuitions representing induction, to understand the interaction which takes place between operator and artifact sufficiently to represent in any code what happens in an experiment; and so on. That is why again the formalization in artificial languages in all their simplicity can hardly be of much assistance in dissecting even the early stages of science. In the later stages the situation is perhaps somewhat different, for in them any aid to the understanding of how mathematical systems are composed and interpreted concretely is valuable. It should be borne in mind, however, that the scientific method is a method of discovery and there can be no methodology of the sciences which does not move in that direction; yet for the purposes of discovery formalization is not very helpful; indeed it may act as a hindrance by indicating a supposed consistency or a premature completeness.

An empirical language rests on the disclosures of experience, and the mathematical language on the disclosures of logic, while the artificial languages have neither the concreteness of the one nor the abstractness of the other. Scientific description has to be constructed in the scientific language. And the scientific language by dialectically weaving together the relations of substance (i.e. matter and/or energy) within the limits prescribed by logic (or of empirical elements as prescribed by mathematics) is saved from excesses of both nominalism and idealism; saved from nominalism by the incurably Platonic nature of all abstractions, empirical as well as mathematical; and saved from idealism by the incurably Aristotelian character of all substances. In the first case, if the names for classes denoted merely the members of the class and not the class itself, classes could not be manipulated in the absence of their members, on the grounds that by so doing we would be engaged in misrepresentation, for we would be guilty of assuming that classes extend beyond their members in a way in which the members considered in themselves did not warrant. In the second place, to suppose that classes, in being different from members, altogether represent their members, would go against the evidence that empirical objects have the common property of resistance to change, and that they enjoy the singularity of temporal and spatial occupancy. Only so long as the classifications accepted as significant by the scientists are reflected in the languages constructed by the mathematicians will the latter be useful to the former. Only so long as the formalizations accepted as valid by the mathematicians are employed by the scientists as their language will the former be useful to the latter.

The scientific language is a communication system whose elements were suggested by, and employed in aid of, the human inquiry into the laws and behavior of material objects. What keeps science empirical is its ability to refer any formula back to some protocol language which in turn rests on ostensive definitions. Such ostensive definitions occur at that point where activity turns over into language. Ostensive definitions cannot be manipulated, and so the investigator soon moves on to an abstract language which can. But empiricism demands that whenever called upon it shall be possible for him to run the film backward from language to activity and in this way to return to that primitive starting stage at which ostensive definition connected them.

(d) *Logical induction.* Induction is of course a logical term. It will be employed here only in so far as it applies to a step in the scientific method. Induction in science has a somewhat restricted meaning. An induction in science, then, is a general proposition in the form of a question, as suggested by particular data. The question itself proposes a generalization. Induction endeavors to secure the elicitation of truth from the disclosures of experience. If the question cannot be answered, or if it is answered in the negative, then it is abandoned, and new data inspected in the hope that new questions will be suggested. If the data themselves were provocative, then questions suggested by them and answered in the negative must be replaced by other questions suggested by the same data: provocative data must be explained. Thus the self-corrective nature of induction in the scientific method is a feature which is peculiar to it and immensely valuable. The successive repetitions of the inductive process is more capable than any other method of approximating closer to usable general propositions. But this method is apart from its character as a single act. The act of induction is an act of consilience; disparate things or events suddenly related through their similarity are made to "jump together." Inductions say that certain items have the character of relatedness but do not undertake to explain why.

A common misconception of induction has been due to its confusion with probability. This is understandable enough, for there is a probability aspect to inductions, and all statistical arguments are inductive in character. Induction is therefore related to probability, but still must be properly distinguished. Not all inductive logic is probabilistic logic. Too often the important inductions which have led to hypotheses which are subsequently verified by observation, experiment, calculation, prediction and control, were suggested by single instance, such as, for example, psycho-analysis, which was suggested to Freud by a single case of conversion hysteria, and relativity theory, which was suggested to Einstein by particular difficulties confronting the Newtonian system of mechanics. Induction is concerned with the discovery and definition of concepts. Probability is concerned with verification or with prediction. Probability-as-verification will be treated in Chapter V and probability-as-prediction in Chapter VI. Probability has to do with the *testing* of an induction, not with its *discovery*. The probability of an induction is increased by the repetition of sampling (Bayes' theorem). But the original induction is or is not correct. Not all inductions lead to formulations of uncertainty. Indeed the so-called logical induction does not.

These men are from Africa,
These men are pygmies,
Maybe all men from Africa are pygmies.

Induction argues from particular instances to a general rule, and the general rule can be, as it is in this case, a universal one. The uncertainty is a function of the *assertibility* of the conclusion, not of the *content*. It is possible to construct an inductive argument which moves to a conclusion from a premise containing a single instance, as in the apocryphal story of Newton who is supposed to have discovered gravitation when an apple fell on his head while he was sitting under a tree, or the more reliable account of the suggestion of the polarization of light from the observations of sunlight falling on a crystal of Iceland spar while it was being slowly rotated. The induction consists in the way in which the conclusion is inferred from the premises, probability considerations are introduced in this example by estimates of the extent to which these pygmies are a fair sample of African men. Induction is a species of argument; probability in this connection has to do with estimates of its truth. The probable induction moves from grounds of relative frequency to a conclusion.

> The percentage of deaths from lung cancer among cigarette smokers is .6 higher than among non-smokers,
> Therefore perhaps cigarette smoking is a cause of lung cancer.

Deduction is drawing a conclusion about a class from a study of its relation to other classes. Induction is drawing a conclusion about a class from a study of its members. But the similarity between induction and deduction which is featured traditionally by casting them both in the form of a syllogism is partly illustrative but also partly misleading. Inductions are related to deductions in a way which has not been entirely explored. For in this connection, induction is the choosing of the premises which are necessary for deduction. (These premises are also described empirically as hypotheses and mathematically as axiom-sets.) Given an abstract deductive structure, such as a mathematical system, induction and deduction become directions, upward for induction, downward for deduction. Deduction means descending the ladder slowly enough to count the rungs. Induction means ascending the ladder so swiftly that the top platform alone is seen. But in both cases it is the same ladder. In the ordering of scientific material there is hardly a stage which does not require both processes.

A special case of induction is the movenment from particular phenomena to particular cause which Peirce termed "abduction." It may be illustrative to compare a probable induction, a logical induction and an abduction.

> *Probable induction:* A relation is found to be the case on more than one occasion. A prediction is made that the same relations will probably be found to be the case on some other occasions.
> *Logical induction:* A relation is found to be the case on one or more occasions. A general proposition is inferred which asserts that such a relation is necessary in some or all such cases.
> *Abduction:* A relation is found to be the case in the present. A particular proposition is inferred that certain events in the past were the cause.

Examples of probable and logical induction have already been given. An example of abduction might be the following. Chipped stone tools were found with the skeleton of an apelike man which, by carbon 14 dating methods, proves to be some 1,750,000 years old. It is inferred that apelike men made and used stone tools some 1,750,000 years ago.

There is no such thing as an immediate inference from the individual facts of the past to those of the future. Such an inference must be mediated by a general proposition. The general proposition may in argument be suppressed, but the resultant enthymene still depends upon it.

(e) *The justification of induction.* Empiricism, which lies at the foundations of the scientific method, requires that the test of truth consists in the data disclosed by experience; yet all arguments based on experience – all inductions – lack logical necessity. Thus it comes about that the search for a justification of induction has been the concern of philosophers for a very long time. Attempts to construct a formal logic of induction have encountered difficulties. Induction seems formalizable in any important way only on the basis of the mathematics of probability. But such a discipline is unable to provide for the induction from a single instance, the arrow that goes straight to its mark. Since so many of the great discoveries in science are of this character, the failure is a serious one. Statistical inductions constitute only one species of inductions, and they may be outweighed by the intuitions of men of genius in the sciences.

There can be no justification of induction which has regard merely for the induction itself, since the result may be a generalization which will not stand up under testing and which is therefore abandoned. The justification of induction cannot rely upon the *act* merely because of its non-demonstrative nature. It must rely, then, chiefly upon the success of its *results*. Science is not built on facts alone but on theoretical systems which function as interpretations of facts. Unexplained facts indicate that a theory is needed to account for them. More specifically, it would not be possible to predict the future from the data of the past and present without some intervening theoretical considerations, and such considerations are usually the result of the operation of inductive inferences.

Induction operates to predict the future from information gathered in the past and present. If a sufficient number of inductions lead to hypotheses which survive to become theories and laws under experimental testing, or if they lead to the choice of axioms from which fruitful theorems are found to flow, then induction is to that extent justified. The leap from material objects to formal statements finds the contents of those statements in an abstract world where they can reinforce their claims by the demonstration of their consistency with other formal statements which have already been established by experimental testing.

Laws of nature are concerned with infinite populations. Finite populations can be taken only as collective instances or particulars, where the collective itself is identical with the class. Thus complete or perfect induction is ruled out in scientific method, for in observation or experiment the investigator is always sampling infinite populations. The sampling of finite populations also produces probable inferences but the sampling of infinte populations is in the interest of discovering or establishing general propositions.

When these are stated as equations, the universality become clear. $y = (f)\ x$ certainly says nothing about the limitations of the two classes, y and x. The universal is an extreme case of the general, and while it is not necessarily to be assumed in the case of a class the extent of whose membership is unspecified, neither is it to be excluded.

The universality which underlies induction as definitely as it underlies deduction is guarded at both ends. It is guarded at one end by the recurrence of similarities among natural phenomena. This corresponds closely to what Mill described as "the uniformity of nature", though the two conceptions are not precisely the same. The recurrence of similarities, like Mill's principle, certainly does underlie faith in the probability of induction. If there were no similarity to events, science would become impossible. It is not the events themselves which recur but certain features which they have in common with other events. It is the similarity which recurs and on which we base our observations and inductively derive and confirm our scientific hypotheses, theories and laws. It is guarded at the other end by the dependence of data upon the prior principle that all of the information derived from experience belongs to a system for which the scientists are looking.

The final justification for induction is the profusion of hypotheses it develops. Inductions are usually made many times, and this self-corrective process turns up hypotheses by the score. The vast majority of these are discarded after cursory inspection, others after considerable scrutiny, while a few only are retained long enough for a testing by experiments and fewer still found worthy of adoption. Theories are already at work in the discovery of hypotheses, and all theory in science looks forward to experiments which shall be crucial in determining which hypotheses are worthy of submitting to further testing.

(f) *Induction to hypotheses.* We have noted that after the observation of data induction is preceded by the construction of classes. From observation to hypotheses, the method of science consists in efforts to move away from the concrete and toward the abstract, and always in abstract terms which can be shown to square with the concrete. Inconsistency with concrete facts has to be eliminated in viable abstractions, for their existence while separate is predicated upon agreement: we save only those abstractions which remain in agreement with observations.

The two best known methods for arriving at abstract systems of knowledge are those which have been described by Wittgenstein and Hilbert. We had occasion to look at the Wittgenstein method earlier in this chapter (Sec. 1, (a)). Now let us look very quickly at the Hilbert method.

Hilbert's contribution has been to add that the discovery and manipulation of abstract systems can be interpreted as games played according to certain logical rules, a method which provides the necessary detachment. Undefined terms are combined into unproved propositions called axioms, from which according to established rules of inference theorems are deduced. Hilbert's abstract structures are purely formal calculi, entirely devoid of content.

The task of combining Wittgenstein and Hilbert consists in filling the empty Hilbert formalism with concrete denotative meanings which have been constructed by means of the Wittgenstein technique.

We can see if we look more closely at Wittgenstein's suggestions the formal structure of the method of empirical science; and at Hilbert's that of mathematics. A fully developed empirical mathematical science employs both methods, each somewhat modified and conducted in stages. Induction, taking off from the ground prepared by construction, furnishes the connecting link between Wittgenstein constructions and Hilbertian deductions. The complex propositions which are the end results of the Wittgenstein technique are related and found to fit the abstract systems deduced in accordance with the Hilbert formula. This fitting is a nice business and comprises the highest stages of science. The procedure is complicated, for further inductions are required to perceive the correspondence. For instance, a group of empirical measurements are to be associated with the symbols employed in the mathematical equations. A selection has to be made in each empirical case from among a bewildering array of mathematical systems. The relations between the empirical quantities will be obtainable from the solutions of the mathematical equations.

It is the *repetition* of induction to hypotheses which distinguishes this stage from the next one in which hypotheses are adopted. What the investigator hopes for from induction is the discovery of an hypothesis which if adopted would yield by deduction the greatest degree of agreement with the disclosures of experiment. Tremendous inductions are required to utilize empirical data in the construction of an hypothesis, and inductions of this sort perform a systematizing service.

It would be a mistake to leave this part of the account of the scientific method with the impression that filling a Hilbert formalism with Wittgensteinian constructions is an automatic method. I have endeavored only to lay open for inspection the bare bones of the method. In actual practice more often than not the process is considerably more random than what I have described. The investigating scientist both observes and thinks, he engages in irrelevant physical effort in order to fuse what he has observed with his thoughts about it, he may take a walk, talk about other things with students or colleagues, glance at samples of the relevant literature, until something significant occurs to him.

Inductions, then, serve to connect constructions with deductions by moving from construction to the discovery of hypotheses which if confirmed by experiment furnish the axioms for an empirically interpreted mathematical system. And so it is to hypotheses that we shall have next to turn our attention.

CHAPTER IV

THE ADOPTION OF AN HYPOTHESES

We have seen that much spade-work has to be done before that stage of scientific procedure at which hypotheses enter the picture. There are no doubt many instances of experiments conducted without benefit of hypothesis. Yet it is only with the hypothesis itself that a line of investigation can be gotten under way. We shall study the character of hypotheses under the following headings: (1) *definition and description*, (2) *character*, (3) *criteria*, (4) *kinds*, (5) *occasions*, (6) *discovery*, (7) *function*, (8) *indispensability*, and (9) *adoption*.

1. *Definition and description*

An hypothesis is a proposition which seems to explain observed facts and whose truth is assumed tentatively for purposes of investigation, an explanatory statement accorded only that degree of acceptance sufficient to justify its further study. We have spoken in Chapter II of hypotheses being employed in *observation*. Here we shall talk about hypotheses employed in *explanation*, which is a far more complex affair. An hypothesis is something a scientist hopes is true, a provisionally accepted theory, a sort of leading question put to nature and intended to solve a problem, a kind of guess specifically designed to suggest the sort of inquiry by which an answer might be reached.

It is important to distinguish at the outset between the hypotheses of science and those of ordinary common sense. Both are attempts to explain phenomena; only, in the case of the scientific hypothesis there is the combination of data extending beyond ordinary experience with inference to generality. The man who says, "This ring was sold to me as platinum, but must be silver because it has begun to tarnish," has proposed an hypothesis arising in ordinary experience which is susceptible of investigation. A corresponding scientific hypothesis in sociology might read, "The tendency to deception in social relations is inversely proportional to the risk involved."

The success of the method of hypothesis depends upon the delicate balance between stochastic and empirical elements. We have to construct a guess as to how things are before we can make the observations necessary in order to determine whether our guesses were correct. An hypothesis is a guess at an answer, and for this there must have been an anterior question. The pointed question contains the hidden hypothesis. The hypothesis, then, is a proposition suspected of being true but lacking the requisite support of evidence, a suggestion that new and necessary relations may exist and that their existence should be investigated. A scientific hypothesis is one that offers a possible explanation or that ascribes an adequate cause. When physicists sought an explanation of the puzzling photo-electric effect, Einstein in 1905 proposed the hypothesis that rays of light consisting of corpuscles carried an energy that varied inversely with the wave-length. What he then described as light-quanta are now called photons. He was able to show that from this hypothesis the photo-electric effect followed as a consequence.

Every hypothesis is a proposition, and as such involves its own truth, an implication presupposing the being of an objective set of conditions which the proposition attempts to represent but which it may or may not correspond. (In the case where the hypothesis is a question, it prescribes its own range of relevance with the proponderance implicit in favor of a correspondence.) The investigator is predisposed to its acceptance but not certain, and more disposed still to the acceptance of a truth which is independent of his dispositions. An hypothesis, in other words, is a preconceived idea as to what an investigation will disclose, maintained on the basis of a certain measure of doubt. It must be a matter about which there is both doubt and some inclination toward belief. Where there is no doubt concerning the truth of the idea, then there will be no willingness to submit it to the test of experiment. No one would investigate a proposition believed on conclusive evidence to be absolutely false, and no one would need to investigate a proposition about whose truth there was no doubt. Without the inclination to belief, there would be no disposition to look into the degree of truth of an hypothesis. The inclination to belief is thus a product of the promise of explanation: if the hypothesis should be confirmed, it could be used as an instrument in further inquiry. And without the large element of doubt, the proposition would be a simple matter of belief, not a scientific hypothesis.

2. *Character*

Every theory or law in science must have begun its career as an hypothesis.

Something problematical in the subject-matter acts as a compulsion to the framing of an hypothesis as a possible solution, and the hypothesis will usually have the property of conceivable testability.

The intuitive discovery of an hypothesis means logically the inductive movement from the provocative facts, which are individual, to some principle of explanation, which is general. Whenever a scientist arrives at an hypothesis, he has got more or less away from the facts. His hypothesis may or may not have been suggested by facts, and it may be tested by facts; but it always goes beyond the facts which suggested it and even beyond those which are used to test it. And this is true even though it refers to facts. But because of the element of generality which is always involved in an hypothesis, there is of necessity something in it which was not given fully by the facts.

The hypothesis is a stage beyond induction also because it relates data which if left to induction might have seemed more disparate; it reaches toward cause in a way which induction does not. The hypothesis has an internal as well as an external aspect. As internal the hypothesis is just its own nature: a proposition neither true nor false but simply an asserted relation stating something concerning all relevant facts or conditions. As external it is the same proposition but poised looking forward to its own verification or falsification. Hypotheses in the main are faced forward toward their ordeals by experiment and calculation, by prediction and control; yet they face backward also toward the observations and inductions from whence they sprang. The hypothesis, undertaken in the first place only with the implicit understanding that if able to remain unrefuted it will lead to further inquiry and perhaps, after the final test, to its own establishment, has a leading edge and a forward motion which lend to it the character of a conditional proposition, undecided as to whether its nature is contrary-to-fact or not contrary-to-fact, and of a question for which an effective answer is momentarily expected.

Thus in order to determine the character of an hypothesis it is necessary only to read back from a theory or law to a more tentative and probative formulation of the same. Of course the formulation in the course of such a procedure may not have remained identical; the hypothesis may have undergone revision in the direction of greater precision, it may have achieved mathematical expression; but what has changed radically as the hypothesis has moved from the status of hypothesis to that of theory and then to that of law is the increased support it receives. The experimental and mathematical evidence will have accumulated, with the resultant strengthening in the claim to truth which it can sustain. If the formulation is not identical, this will usually be found to mean that as it has moved along it has been given more precise and more abstract expression, usually of a mathematical nature.

There are weak and strong hypotheses. A weak hypothesis is one where little is known concerning the circumstances surrounding it. A strong hypothesis is one where the initial conditions and the intervening laws of change are well known but the extenuating circumstances are not known, so that what otherwise might have been a causal law must remain something of an hypothesis. It should be added that weak hypotheses often receive support and are continued as laws; their weakness is not anything resulting from a tendency to error. An example of a weak hypothesis is that concerning the effect of solar prominences on radio communication blackouts. Between these extreme types lie the bulk of scientific hypotheses. The ideal hypothesis would be one expressed in sufficient detail so that a decisive experiment would be readily suggested by it.

Hypotheses are empirical propositions which aspire to the condition of logical propositions whose reference lies beyond the limits of the knowledge of them and whose status is independent of the actual world in which they are illustrated. But aspiration is not achievement; and meanwhile the hypothesis is a proposition in a temporary state; it is an explanation on probation. It may have been developed quickly or slowly. But no hypothesis is intended to remain an hypothesis permanently; it is not a proposition occupying a slot in a science which it can keep forever. An hypothesis is a proposition on trial, and it will remain on trial only so long as is necessary in order to verify or falsify it. The scientists call it a "working hypothesis." N. P. Krenke for instance suggested that since the leaves of some varieties of cotton change shape as they occur higher in the stem, this difference could be used as an index to the physiological age of the plant. Before the positron was found experimentally in 1931 by Carl D. Anderson, Dirac's prediction the year before that it would be found had for that year the status of an hypothesis.

Eventually, after sufficient study the working hypothesis will be either elevated to the status of theory and passed along to another type of consideration, or dismissed as false and unworthy of further consideration in any connection. The condition of hypothesis is a means to two conditions which are not those of hypothesis: a discredited false proposition or a sustained theory.

In addition to being tentative, hypotheses are heuristic, and this in five ways: (a) hypotheses may suppose generalizations not yet tested or otherwise investigated; (b) they may suppose entities not yet observed or isolated; (c) they may suppose processes not yet observed or isolated; (d) they may suppose properties not yet observed or isolated; and (e) they may posit ideals in terms of which specific inquiries can be conducted. This classification of hypotheses is rough indeed and a certain amount of overlapping is unavoidable.

(*a*) The first is the kind of hypothesis we generally have in mind when we think of hypotheses, and one we have been discussing thus far. An hypothesis of this type is usually a statement about a universal relation, expressed perhaps as the statement of an invariant relationship between variables or as the description of a set of conditions. The first proposal that perhaps epithelial tissue is constituted chiefly of keratins did not mean that insoluble proteins of the scleroprotein class, containing large amount of leucine, cystine and tyrosine radicals, were to be found in *some* instances in the analysis of the epidermal layers of skin; it meant rather that in *all* epidermal layers of skin and in *all* horn, hair, nails, feathers and hooves the keratins are the chief ingredients.

It is a fact that we are concerned with the degree of truth of our hypothesis considered as a general proposition capable of being elevated to the status of theory and perhaps also to that of law. But we are concerned also with a larger area which our generalization may help to explain. To say then that a generalization is heuristic means that it furthers explanation, that it serves to elicit other generalizations or facts which without its aid we might never have reached. An hypothesis is a guide to a fresh viewpoint, to novel observation, and to original experiments.

(*b*) The second of our ways is also to be included in the way of hypothesis: to suppose entities not yet observed or isolated. Hypothetically proposed or heuristic entities are quite common in science; we find them in physics among the sub-atomic particles, many of which were inferred before they were experimentally demonstrated. Dirac predicted the discovery of the positive electron (the positron), and in 1935 Yukawa predicted the meson. Both entities when announced were of the nature of hypotheses. The discovery of the anti-proton has led to further speculation in the same direction – to the supposition, for instance, that corresponding to every physical particle there may be an anti-particle, and that there may be anti-galaxies and even such a thing as anti-matter. Anti-protons and anti-neutrons were, in fact, proposed as hypotheses and later produced experimentally as a result of investigations conducted with the Bevatron accelerator at the University of California. On the other hand, the diseases of psychopathology, paranoia and schizophrenia for instance, have never been observed; but they have aided in the understanding of the unconscious.

(*c*) What is true of entities is also true of processes: they may be hypothetically proposed and unobserved, yet a great aid to explanation. The biological process of maturation and the economic operation of "conspicuous waste," are of such nature. The Lorentz-Fitzgerald contraction was an hypothesis calling for the shortening of length and the slowing of clocks at very high speeds. The proposal made by Uhlenbeck and Goudsmit in 1925 that electrons possess spin is another case in point. Inaccessible processes resemble heuristic entities in that the transformation of an hypothesis into a theory or the justification for the abandonment of the hypothesis often has to be postponed pending the discovery of new techniques.

(*d*) Hypotheses may suppose properties not yet observed or isolated; yet they may when investigated lead to new knowledge. William Gilbert in the sixteenth century experimented with magnets and observed that a freely suspended magnetic needle pointed north and south in directions roughly comparable to the geographical poles and that the angle of the magnetic needle when pointed toward the north pole dipped through an angle that varied with the latitude. From these facts he made an induction to the hypothesis that the earth had the property of an enormous magnet.

(*e*) The statement of hypotheses as ideals is a familiar one in the sciences. "Ideals" in the scientific usage does not connote anything peculiarly mental; they are abstract. Patterns of perfection are necessary limiting cases; they are models and nothing more. They serve an operational purpose. It often seems to be necessary to posit ideals in order to investigate actualities; for then standards are provided for measuring degrees of approximation in actual cases. Carnot's conception in 1824 of a theoretically perfect engine in which there is no loss of heat by conduction and no loss of work by friction is one of the best known examples. An ideal elastic would rebound with undiminished speed after colliding with the walls of its container. The equation for an ideal gas calls for an average distribution of molecules. The ideal "black body" of thermodynamics is the hypothetically perfect heat radiator. The ideally smooth plane has a static molecular surface.

One important characteristic of hypotheses is that they usually extend beyond ordinary observation. This is accomplished with instruments, by means of an extension of the senses, or with mathematics, by means of abstract symbolic functions. Thus, while hypotheses start from common sense levels and may have been suggested by observations which were also of that character, they do not themselves directly refer to those levels but instead to conditions far removed from it. From the Ptolemaic hypothesis to the Copernican to the Keplarian to the Newtonian to the Einsteinian, there is a continual increase in degree of abstractness even though all are attempts to account for much the same kind of phenomena (although, it is true, not the same range). Scientific observation certainly does begin at the level of ordinary common sense but its eventual aim is perfectly abstract mathematical formulation: the representation of empirical statements in tautological form; and the hypothesis is the first step in this direction: it takes off from the ground of familiar data, perhaps, but it does get off the ground.

Hypotheses, like their more established counterparts, the theories and laws, extend beyong the mere summaries of relevant data. That for almost all gases the pressure times the volume is a constant, (Boyle's Law) means that there is a generality involved in every specific instance which extends beyond that instance to others which are similar. The equation, in other words, goes beyond the data even though clearly suggested by observational data and partially confirmed by experimental data. It is for this reason that the hypothesis, unlike the observation or the experiment is a logical affair and a candidate for the status of theory and law.

It is usually true that the greater the number of hypotheses the better. Any hypothesis is better than none if it leads to an investigation that would not otherwise be made, for experiments conducted to test one hypothesis may frequently suggest another one which is more satisfactory. Hypotheses rarely occur singly, although in order to study their character we have been discussing them as single generalizations or single entities. Since hypotheses are discarded so rapidly, every scientist must have at his command an abundance of hypotheses. And since they must account for the same phenomena and must share areas of common agreement, they constitute families of explanation. Nests of hypotheses may occur to a single investigator or they may be collectively brought to bear on a problem by a group of investigators. In either case, rival explanations of the same crucial phenomena have to be tested and those found wanting discarded. For every hypothesis that passes on to the theory stage, many are abandoned as inadequate. The life span for an hypothesis may vary from the flicker of an instant to many years. An hypothesis may occur to an investigator in the process of sorting out possible explanations and, because of obvious limitations and inadequacies, be instantly given up, or one may have been announced by him for its explanatory suggestiveness and then because of the difficulties confronting its investigation, await for years the decision regarding its final disposition.

3. *Criteria*

We have next to consider what are the proper criteria of good hypotheses. What will influence an investigator to make a choice among rival hypotheses, to choose one which seems better to him than the others? We have said already in an earlier section that a good hypothesis is one that offers a possible explanation. It is necessary now to be somewhat more specific and to break this requirement down into six lesser but more precisely stated requirements.

(*a*) A good hypothesis is one that can adequately explain the observed facts. No one would ever examine an hypothesis without this hope, for any which is found to be false will have to be abandoned and the work already done on it written off as wasted. An hypothesis for which data can be claimed in its support is already well on its way to becoming a scientific theory or a law. The very fact that it has been chosen means that it has passed the first rudimentary test and will now be handed on to the second, which is the test by experiment; for the choice indicates that it is preferred over many others.

The hypothesis of gravitation, suggested to Newton in 1665 by the fall of small objects toward the earth, and extended by him to account for planets remaining in their orbits, is the classic example. Obviously, if the problem which gave rise to the hypothesis consisted of the uncovering of crucial or provocative facts which had gone unexplained, the hypothesis would be cast in the form of a principle of explanation or alternatively in some form from which it might be possible to deduce a principle of explanation.

(*b*) A good hypothesis is one that offers an explanation which is the simplest possible under the circumstances. A modified form of Occam's Razor is applicable here. Of two hypotheses which can equally account for the data under consideration, the simpler is to be preferred. Economy of explanation is one of the guiding principles of science. The Copernican hypothesis to account for the motion of observed planets is simpler than the Ptolemaic, although both are equally authentic scientific hypotheses. Einstein's theory of relativity was chosen because it is simpler than Whitehead's.

(*c*) A good hypothesis is one that offers an explanation which is as complex as necessary under the circumstances. This is the counter-principle of Occam's Razor, and though is has no name, it is to be found in Kant's *Critique of Pure Reason*, A656. It could be called Kant's Shaving Bowl perhaps. It is simply the demand for adequacy of explanation: a good hypothesis will be broad enough to explain all of the observed facts. Gibbs' phase rule in chemistry is a formula for classifying all heterogeneous systems in equilibrium. Its complexity has proved adequate. However, a qualification must be noted here. All abstract formulas are stated as general propositions which are adequate to cover an infinite population of instances, and thus in the very act of meeting the requirements of adequacy to explain the observed facts – always of necessity a finite number – they go far beyond them.

(*d*) A good hypothesis is one that is conceivably verifiable, that can be brought into agreement or disagreement with observations. It must be capable of deductive development into consequences that can be verified or falsified by the data disclosed to experience. In other words, it must be cast in such a form that it could conceivably lend itself to investigation. For example, a sample of liquid is given to a chemist to examine. He notes that it is a crystalline solid, dark grey in color and shiny. He guesses it to be iodine. This is a stochastic hypothesis. Upon examination he finds that it is not very soluble in water but that it is soluble in a solution of potassium iodide and that it turns starch blue. He concludes that his hypothesis has been confirmed. But on the other hand a theory that the meta-galaxy has outer limits would be difficult and perhaps impossible to verify.

A hypothesis in science must keep to predictions can be observed or experimentally verified. Verifiability is difficult to define and is a property which an hypothesis may have or not have, gain or not gain. Verifiability changes as techniques improve. Two decades ago the proposition that there are craters on the far side of the moon was cited as a good example of an unverifiable proposition, as indeed it was then. Since then it has been verified by moon landings of instruments and men.

Verifiability is the name of a range which includes falsifiability. If the latter is not possible, the former is not meaningful. Any hypothesis in science which cannot be falsified cannot be said to be verified in any important way. Popper is correct in insisting that whatever is insusceptible of falsification cannot be meaningfully verified, and that practicing scientists deal only with refutable hypotheses, but he is wrong in supposing that verifiability is thereby ruled out. The failure of falsification is not the only criterion for verification. Certainty can be affirmatively approached.

Hypotheses which are referable to an empirical subject-matter but which are by their nature insusceptible of verification have no place in science. But impossibility of verification is difficult to demonstrate, and changes from time to time. The proposition that there are craters on the other side of the moon was not so long ago a favorite example of unverifiable hypotheses; but since the lunar program of the United States' National Aeronautics and Space Administration, the far side of the moon has been not only seen but photographed, and the proposition has passed into the class of verified hypotheses.

It is not always possible to tell by inspection whether or not an hypothesis meets the qualification of conceivable verification. As our equipment improves and more of the environment is brought within the reach of experiment the conception of what is and what is not verifiable changes. It would never have been dreamed a hundred years ago that such an hypothesis as that of Planck, that the energy of an oscillator of natural frequency ν is $E = nh\nu$, where n is any integer from 0 to ∞. Yet is was not long before the observed distribution of radiation confirmed it with an allowable degree of experimental error.

Although the investigator does not always know when an hypothesis is conceivably verifiable, he does sometimes know when one is not. An hypothesis is not conceivably verifiable when it is a metaphysical hypothesis. Metaphysical propositions are necessary as assumptions in scientific method, but for that very reason they are illicit when introduced as elements in its operation. Those who oppose metaphysical statements in experimental science are correct in their opposition provided they do not carry it to the extreme of asserting that the scientific method as a whole makes no assumptions whatever concerning metaphysics.

(*e*) A good hypothesis is one that is strong enough to compel inquiry. It must be sufficiently intriguing to urge the investigator to try to verify it. And he will try if it is the kind of hypothesis that carries a high explanatory potential or that, once verified, might suggest many further experiments. It would be difficult indeed to deny such a claim to Darwin's original hypothesis of the origin of species by natural selection. Some hypotheses are so provocative that they open up entirely new lines of investigation, and it is these hypotheses which are apt to be tested before others.

(*f*) A good hypothesis is one that is sufficiently abstract. The prejudice of working scientists in favor of the concrete should never cause them to lose sight of the fact that in the laboratory or the field, with instruments or material techniques, what the scientist is testing is the validity of an abstraction. Abstractions not in some relation to the concrete are not scientific abstractions; but on the other hand statements insufficiently abstract are apt to be those of common sense and of ordinary experience rather than of science. In the case of useful and familiar abstractions, simple arithmetic and Euclidean geometry for example, we tend to forget their abstractive nature: familiarity and use have tended to lend them a concreteness they do not in fact possess. The more independent an abstraction is from the material world, always provided that it can be kept in some relation to that world, the more we are able to control matter by its aid.

4. *Kinds*

It should be emphasized that with respect to definition and function there is only one kind of hypothesis. The kinds are distinguished not qualitatively but rather quantitatively and structurally, and with respect to the variety of data encompassed. In general, hypotheses assign causes to effects by positing mechanisms; but there is a degree of complexity involved which is proportional to the variety of kinds.

Hypotheses of first order undertake to account for the relevant data by asserting them to be the effects of supposed causes. Facts unknown are hypothesized as causes of facts known. When a photographic plate was left by Madame Curie in a drawer next to a piece of radium and was found later to have been exposed, it was hypothesized that the radium had developed the plate.

Hypotheses of second order undertake to account for the relevant data by placing them under an over-arching generalization. Newton's hypothesis of gravitation is a good example. Hypotheses of second order are of necessity more inclusive and more abstract than hypotheses of first order. There is here a higher grade of complexity which insures that both shall be the case.

Hypotheses of third order undertake to account for the relevant data by assuming that what is true of one set of data will be equally true of another. Maxwell guessed that what had been found true of gravitation would be found equally true of electrostatics, that the inverse square law of gravitational attraction could be found to hold for electrical attraction; and the result was the electromagnetic explanation of light which had such brilliant confirmations in both magnetics and electricity. The hypotheses of third order are apt to be more mathematical than the others, as the only way in which the proper degree of abstraction and generalization can be made available. Universality is a function of complete abstraction; and although complete abstraction can never be fully obtained, the exigencies of communication systems being what they are, it can be approximated.

A special kind of hypothesis is the *ad hoc* hypothesis. The *ad hoc* hypothesis may account for one recalcitrant fact, or, more commonly, it may be needed to fill out the gaps between theories which have been successfully tested. In the latter case, it may or may not be conceivably testable. The discovery of Neptune was due to observed irregularities in the orbit of Uranus as called for by Newtonian mechanics, irregularities which were explained by introducting the *ad hoc* hypothesis of the existence of another planet – in this case later verified. The hypothesis of a luminiferous ether which could offer resistance to motion was posited *ad hoc* to account for the discrepancy between calculations and observations in the case of the periodic return of Encke's comet – in this case never verified.

We have been speaking of causally framed hypotheses, but there are also statistically framed hypotheses. A statistical hypothesis is a proposition in statistical form having the status of an hypothesis. In general statistics are used in the verification or falsification of an hypothesis, not in the framing of it. But there are special cases where the hypothesis itself is expressed in statistics. We shall have more to say about this kind of hypothesis in Chapter V.

5. *Occasions*

We have reviewed some of the criteria of a good hypothesis. We shall now consider when they arise. Hypotheses are never introduced *ad hoc* but always for reasons. However, quite diverse occasions function very well as reasons. We shall consider five of them.

(a) Perhaps the most frequent occasion for considering hypotheses lies along the path of the regular development of the scientific method between first-order construction and induction on the one hand and experiment on the other. When it is first noted that the induction from first-order construction can be faced in another direction, that is to say toward experiment, then the induction is placed in the position of an hypothesis and is henceforth regarded as possibly true and testable. The hypothesis already existed in objective propositional form as the result of an induction; but now it is to be regarded as a conjecture. When Darwin came to examine the origin of species, he had already at hand the first-order constructions of the taxonomists and the artificial selection practiced by the breeders of poultry and other domestic animals. His own conception of natural selection was an induction, and only afterwards treated as an hypothesis.

(b) Quite commonly an hypothesis is developed when problems are encountered which cannot be solved on the basis of existing theories, and unexpected data are turned up which the old established laws cannot explain. Harvey discovered that the heart deals in a minute or two with as much blood as there is in the entire body, which suggested to him that, since the heart keeps on pumping, the blood must somehow travel from the arteries to the veins and so back to the heart. This single, unexplained fact of the volume pumped induced him to discover the hypothesis of the circulation of blood, which he announced in 1628. It was the failure of the Rayleigh-Jeans Law to predict the distribution of radiant energy in a black body that led to the investigations which culminated in Planck's discovery of the quantum constant. Again, the Fitzgerald contractions and the Michelson-Morley experiment could not be accounted for by Newtonian mechanics. A new hypothesis broad enough for both had to be developed.

When rival theories are in conflict, a temporary sort of settlement may be reached through some wider hypothesis. Copernicus was confronted with the established Ptolemaic theory and the theory of Nicetas, who – according to Cicero – thought that the earth moved. He therefore in 1543 proposed the hypothesis that the stars were fixed while the planets in the solar system, including the earth, were in motion. The so-called unified field theories could perhaps more accurately be described as hypotheses were they not already so mathematical by nature and did they not incorporate the results of so much experimental work. A number of these – the theories of Einstein, S. N. Bose, Vaclav Hlavaty, and Heisenberg – in 1960 contended for the unification of physics.

Experiments may seem to disclose conflicting data, and here, too, resolution through the positing of an hypothesis is called for. The evidence for light as a stream of particles seems irreconcilable with the evidence for light as a wave phenomenon.

(c) Hypotheses are devised to bridge from raw data to the explanatory principle which accounts for them. Either may fail of establishment to some degree as a result of reliance upon the other. Either the data are accepted and the explanatory principle is presumptive, which is the usual case; or the principle is accepted and the data are inferred. Young's experiments with interference phenomena at the turn of this century, with the resultant supposition that light must consist of waves, is an example of the former. Freud's principle of an unconscious mechanism moved by the proposed entities of "ego," "super-ego" and "id" exemplifies the latter. Whichever side the established evidence falls on, the other side will always be presumption.

(d) Hypotheses are often discovered by chance. Inadvertent leads often occur to workers immersed in an investigation. Inquiries are dropped in favor of more promising new ones, for the intuitions of great researchers often indicate favorable lines of inquiry to be pursued in the future. Malus's discovery of the polarization of reflected light and Curie's discovery of radium, are among the most famous instances. When Haüy in 1784 accidentally dropped a crystal of calc-spar, he noticed the beautiful geometric faces and understood enough of their importance to drop other specimens deliberately in order to study the regularities. In this way the science of crystallography got its start. Others must have noticed the same phenomenon before without understanding its significance. Galvani touched the leg of a frog with a piece of metal entirely by accident, and it was an accident in a laboratory which precipitated the discovery of the wave-property of matter by Davisson and Germer. Darwin's work on insectivorous plants resulted from a chance observation. Chance discoveries, such as these, are possible only to investigators with a broad deductive background in a particular science. A man in one field of inquiry does not accidentally make a discovery in some other. Prolonged preoccupation with a particular subject-matter seems to be a much overlooked prerequisite of all chance discoveries. In short, the probability of serendipity is a function of the number of observers and experimenters with deductive training and long absorption in the science. The number of unknowns in any situation being what they are, it would be impossible to rule out chance occurrences, and some of these are bound to be happy occasions; the experimenter cannot foresee everything that might happen, and he cannot see, either what might suggest further exploration.

(e) The exploration of practical problems may be the occasion for the discovery of hypotheses in pure science. A significant amount of pure science has come out of industrial laboratories where the solution of problems in applied science was being sought. Much of our knowledge of the half million carbon compounds issues from the search for patentable products. The applications of scientific knowledge to situations involving usefulness calls on a knowledge of science; and expediency, far from driving out principles, requires them. The hypothesis of atomic fusion resulted from efforts to improve the efficacy of the atomic bomb as an instrument of war, and of atomic energy for the practical purposes of a world at peace.

In addition to the above occasions when hypotheses arise, there will seem to be many other occasions when they are not present and do not seem to be needed. These will be when observations are explored by means of various precise methods, such as group theory in quantum mechanics. Newton's contention that he framed no hypotheses could be interpreted to mean that he did not appeal to explanatory principles too far removed from the data, such as for instance metaphysical principles would be. His careful experiments to show by means of the prism that white light contains all colors only made explicit a proposition which must have implicitly suggested his investigations in a given direction. Then, too, in such cases the various, relatively minor consequences of previously established theories (which were once hypotheses but are so no longer) are being worked out. Hypotheses if successful lead to operations and experiments which sometimes do not seem to require further hypotheses. But such a situation soon runs its course in consequences both theoretical and practical, and cannot be continued indefinitely without further stimulation; and so the more familiar rhythm involving occasional hypotheses is eventually resorted.

6. *Discovery*

We have now to say a few words about the mechanism of the discovery of hypotheses. The first thing to remember, perhaps, is that it has not been possible to find any necessary path to the discovery of important hypotheses. No inevitable mechanism is known which, once set in motion, will carry the investigator unfailingly from immersion in the data of his science to the assumption of important hypotheses, even though it is true that the former will often lead to the latter. Experience is needed for the discovery of hypotheses, but it is not a sufficient cause; for there are so many instances of experience which have led investigators nowhere.

Hypotheses may be discovered by induction from individual instances or by induction from more inclusive theory. The former is by far the more familiar procedure of the two. Laboratory experiments are what usually suggest inductions; work at the desk is more likely to produce deductions. Fleming's inference that the micro-organism responsible for the contamination of one of his plates, on which he was studying mutations among staphylococci, might have been an anti-bacterial substance, was an example of the discovery of an hypothesis by induction from the data. The discovery of an hypothesis by deduction from a more inclusive theory is illustrated by the construction of a mathematical model which is yet to be tested against the empirical data. Einstein's unified field theory whereby he sought to bring together gravitation and electromagnetism is an example.

When hypotheses are the results of inductions from observed instances, the mechanism is equivalent to working from particular cases to a general principle or from effects to a cause. Induction founded on probabilities are of course strengthened by increasing the number of instances. But an induction not so founded is not altered in any way. Neither of the two great theories of modern physics: the theory of relativity and the quantum theory, were suggested by statistics or probability. They were inductions from instances of data. A hundred thousand contaminated plates would not have suggested any more to the Curies than the ones did which led to the discovery of radioactivity. An hypothesis induced from one case is no weaker as an hypothesis than the same hypothesis induced from several thousand. Instances do count but only when in the support or verification of an hypothesis, but this is a different, and a later, kind of induction.

The ideal of science is system: every science wishes to put its findings in order, and at present all known types of order are deductive. The role of hypothesis is to suggest the missing axioms from which theorems in agreement with observation can be deduced. An hypothesis and the facts to which it refers stand in the same logical relation as axioms and the theorems which can be deduced from the axioms. In choosing an hypothesis the experimenter is reaching inductively for something which is to stand to the experiments which are to test it in the relation of an axiom to its theorems: he intends to employ it for purposes of deduction.

Accordingly, hypotheses are more often than not inductively discovered; but this does not mean that they are inductive *by nature*. Induction in this connection is the choosing of axioms, which are wanted of course for deductive purposes. An hypothesis is a proposition proposed for consideration, and as such it is neither inductive nor deductive. Induction refers to the manner in which a proposition is suggested by data disclosed to experience; deduction refers to the manner in which a proposition fits into a scheme of propositions. Thus there is no such thing as an hypothesis inherently inductive or one inherently deductive; there are only hypotheses inductively discovered. Although the discovery of hypotheses calls upon that rarest of human faculties: imagination, and the presence of an hypothesis of some sort is the *sine qua non* of the scientific method, it remains true that hypotheses are intended to fit eventually into a deductive scheme or else to be abandoned as worthless.

It often happens that scientific hypotheses are discovered simultaneously by several investigators who may be entirely unknown to each other or so separated by space and time that each is ignorant of the other's work. The simultaneous discovery of the infinitesimal calculus by Leibniz and Newton is the famous instance, but there are many others, such as hyperbolic geometry which was discovered at about the same time by Gauss, Bolyai (father and son) and Lobatchewsky around 1804, or the discovery of natural selection by Darwin and Wallace in 1851, the discovery of the combining power of the carbon molecule (Valency) by Couper and Kelkulé in 1858, or the discovery of the periodic table of the elements independently by Lothar Meyer and Mendeléeff in 1869. Although these are inductive discoveries, the coincidence argues that there is a wide penumbral deductive background in which investigators who are professionals are continually immersed. A science carried to a certain point represents an advance on a broad front, and inquiries conducted in it are productive of hypotheses entirely different in character from those encountered in a science at an early observational or classificatory stage. The influence of the members of the scientific confraternity upon each other is well-marked but far less well-understood. Suffice to say here that the social effects of an advanced science upon inquiry are such as to blur the distinction in use between inductive and deductive procedures.

Hypotheses, then, are usually the results of inductions. More rarely, they may be arrived at deductively, as we have noted above. From a broad theory a deduction may be made to some proposition which can be tested for agreement with observation while the theory itself is receiving mathematical treatment in order to demonstrate its consistency. Thus hypotheses are arrived at deductively further along in the chain of investigation rather than at the beginning. Confirmations of the special theory of relativity continued to be made experimentally in order to test lesser hypotheses long after the mathematical consistency of the general theory had been accepted.

7. *Function*

The purpose of hypotheses is to lead the investigator to new knowledge; the first aim is to move him from suggestions which are uncertain to information which is more reliable. The hypothesis almost as though of its own accord oc- to him as a vague feeling; it formulates itself explicitly in appropriate language; his efforts to dismiss it grow weaker as it becomes more insistent; all other urges and interests are crowded out and his defenses break down completely; the hypothesis now fully in control of the situation demands and receives his full attention as he faces the prospect of designing tests of its significance. Archaeologists excavate stones which can be interpreted as the foundations of a large palace on a site in the southwestern Peloponnesus, and they argue that this may have belonged to King Nestor of Homeric times. The inference from present-day findings to a former state of glory rests on the general proposition that when buildings are destroyed by fire, their stone foundations remain indefinitely. Hypotheses, being general propositions, always extend beyond the data no matter how large the collection, as we have noted above. They are statements about populations indefinitely large and perhaps infinite, and are not limited to any collection however great. For the rule holds that no number of individual instances will ever exhaust a universal proposition, and a universal proposition is one that can be predicated of indefinitely many.

In the case of hypotheses which pertain to individuals it is equally true that such hypotheses are not limited to the individuals. For it is *types* of individuals and not the individuals themselves that are being considered and proposed for investigation. The anatomy of a rat would interest few unless what was true of that rat could be assumed true for most and perhaps all. It is the type the investigator has in mind when he observes a specimen, and not that specimen. He will not concern himself with nor preserve that particular rat, but he will retain instead only what he has learned of its type.

Thus the hypothesis suggests new lines of evidence prompted by the examination of old types, and puts the scientist in the way of imagining relevant types for which he can then devise new experiments.

In this fashion an hypothesis serves as a directive anticipation. It tells the investigator what to look for, and where. It defines a field of data as instances of a generalization, so that the achievement of verification or falsification is rendered feasible: there must after all be something to verify or falsify. It is hoped that what can be deduced from an hypothesis will be in agreement with observations. Mendel's hypothesis that inherited characters remain unchanged certainly altered the direction of investigation once his work was recognized. No scientist ever undertook an experiment without having before him some notion, however vague, of what it was that he hoped to find. Even pilot or exploratory experiments are conducted in order to probe into or run over a range of possibilities, and this constitutes his hypothesis, which is no less one for being implicit or unacknowledged. That he had never formulated clearly and explicitly the assumed hypothesis which led to his investigations is no evidence that it was not there. For why else would he have chosen one experiment rather than another? Implicit hypotheses, operating as assumptions dimly understood, are still hypotheses. Every theory that we entertain and every law upon which we rely was begun at some point in its development as an hypothesis.

Hypotheses do not have to be suggested by observations, although that is the usual procedure in science; but they do have to be verified by experiment, calculation, prediction and control. The fate of most hypotheses of course is to be considered and then rejected. Therefore it is essential to any investigator to have on hand a vast number of hypotheses and to be fertile in their production. Almost any hypothesis is worth entertaining. It is important not to set limits to inquiry. At the same time it is necessary to choose rather carefully those hypotheses which are to be tested by means of experiments since the latter are costly in terms of time, energy and expenditure. This is the point at which the self-corrective nature of the scientific method operates in terms of hypotheses: the successive repetition of the adoption of hypotheses insures that the discarded hypotheses will be replaced by others not yet discarded and possibly worth retaining. An able investigator is one having sufficient imagination to add to his deductive knowledge and technical skill to enable him to produce an abundance of hypotheses. He will admit to consideration all possible ideas, with a view to letting experiment decide. Out of this array, there emerges perhaps one hypothesis which can survive its first test. An important feature of hypotheses is that discredited candidates may turn out to have been of some value. Important discoveries have often been made as a result of the investigations undertaken to explore an hypothesis which later proved to be false. Of this character is the hypothesis of the ether, for its investigation did not disclose the presence of ether but it did contribute to the discovery of the special theory of relativity.

Most questions in science are settled by experiment. But it is the hypotheses that are being settled. Although hypotheses are adopted for the sake of experiment, any attempt to explain the method of science which left out or derogated the role of hypotheses would be similar to an attempt to understand the anatomy and physiology of the human organism by examining only those cadavers from which the heads had first been removed. An hypothesis is a device whereby one set of data with which we are confronted can be connected with another set of similar data yet to be chosen. An experiment is the way in which it has been decided to collect the second set. Thus the hypothesis, however strongly it rests on observation and the inductions made from first-order constructions, looks forward nevertheless to the tests by experiment and calculation, by prediction and control, at which it was aimed.

8. *Indispensability*

Hypotheses are indispensable, but many theorists shy away from the question, leaving the vexed problem of hypotheses to scientists possessed of the requisite imagination, and undertaking the discussion of the method only after the event, on the ground that no rules of induction can be found which will guarantee the discovery of important specimens destined to be recognized as laws through verification by the usual methods of testing. Both Hempel and Popper entertain this view.

The most powerful contemporary school in the philosophy of science has been that of positivism, and the positivists for the most part have dispensed with hypotheses in their account of the scientific method, on the ground that science is descriptive rather than explanatory and the laws of science summatory rather than universal. The issue of positivism is sharply joined over the existence of hypotheses. If they play a necessary part in the scientific method, then the positivist claim of confinement to fact according to the criterion of availability to sensation is denied.

From a strict holding of radical empiricism it would be impossible to discover scientific laws. Individual sense impressions without any generalization to put them together could hardly produce either the theories or the laws with which science is familiar. At the same time, the positivists are correct in their insistence upon the importance of sense experience. It comes to this, that all laws contain hypothetical elements to the extent to which their generality covers more than the reports of sense impressions. The proposition that the electrical field strength in the neighborhood of a point charge is inverse to the square of the distance from it (Coulomb's Law), or the proposition that the hydrogen atom contains exactly one orbiting electron are "facts"; yet they go far beyond sense experience. No number of investigators could claim immediate experience of all electrical fields or all hydrogen atoms, and under the circumstances it must remain impossible for them to do so. There are gravitational fields in inaccessible galaxies and hydrogen clouds far removed from our solar system. Yet pending information Coulomb's Law and the constituents of hydrogen are accepted.

The truth is that under the present scientific arrangements scientific propositions are *answerable* to the facts as disclosed to sense experience but not *confined* to them. The atomic hypothesis in chemistry certainly went beyond the known facts in chemistry when it was first proposed by Dalton early in the nineteenth century; for it assumed the existence of atoms which had not been observed. There were not at the time any facts corresponding to his hypothesis; and if his laws of definite proportions and multiple proportions were later verified by experiment, this means that the facts were found to support the hypothesis. But if the rule had been rigidly followed that hypotheses must be confined to facts, the atomic hypothesis could never have been seriously considered.

Confinement to fact forbids speculation; but the controlled imagination is the kind of experience whereby the generalizations necessary to assemble the facts of a given kind (including those not yet experienced as well as those experienced) are found, and they are called hypotheses. The hypotheses discovered in this way are to be tested in a number of other ways and so held answerable to fact. They were suggested by fact, they will be tested by fact; but due to the generality of their nature they themselves are not factual in the narrow sense of disclosure to sense experience.

The effort of the positivists to eliminate metaphysics from the operations of the scientific method is sound. Metaphysics belongs among the presuppositions of the method, the structure of the method and the interpretations of the conclusions; but it does not belong in the method as one stage of it. Assumptions broad enough to underlie the whole cannot be introduced into the whole as one of its parts. Hypotheses, however, are not necessarily metaphysical; to assume that they are is positivism's mistake. Wave mechanics is not metaphysical, general relativity is not metaphysical, biological evolution is not metaphysical; yet all of these comprehensive theories began as hypotheses and remain, despite the large measure of their establishment, hypothetical in part, that part which extends beyond the testable segments.

Between the fact as observed and the laboratory results there must lie propositions broad enough to subsume the facts and narrow enough to predict the results, and these propositions are called hypotheses. It is a common practice for scientists to employ hypotheses without knowing that they do so. From Bacon to Carnap investigators have supposed that scientists can get along without hypotheses. The error rests on the unobservable nature of the adoption of hypotheses. It is possible to watch a scientist observing or experimenting, and it is even possible to watch him calculating; but it is not possible to watch him adopting an hypothesis. The scientists themselves are as apt others to make this mistake. Poincaré recognized the existence of hypotheses and thought of them as unconscious or natural hypotheses. Newton thought that he got along without them, but he based his optical theory on the hypothesis of emission. What else was gravitation when he first announced it? It can be seen afterwards that both men had bridged from fact to theory by the use of hypotheses. The working hypothesis, so-called, is an instrument of the scientific method and no less so when the scientist is unaware of its silent role. Science with its laws is not the kind of complete system that its relation to sensations alone would indicate. It is a system in the making, and the part played in it by hypotheses shows the character of the method as a process; it is a stage on the way, not a static set of propositions in which the sensations are summed up.

Hypotheses derived from speculation without observation are often called theories; and when they lead to experiment in the same way it is supposed that no hypotheses had been used. Thus theories made up without any immediate empirical support can later be shown to have important experimental results. The atomists of the early nineteenth century had empirical data. De Broglie's wave-equation did not follow an observation. Of course, given an experimental scientist with considerable laboratory experience it is doubtful whether any idea about his own science which occurs to him can be said to lack all observational support. New ideas in chemistry do not occur to politicians or poets but only to chemists. This is not the whole story, however. An hypothesis which is called a theory is no less an hypothesis. If the function exists it justifies the name, for it is not the name which brings about the function. It even happens that an hypothesis is called a theory without having been submitted to any experimental test. This occurs when the hypothesis is complex and can be judged by its predictive or explanatory power. The theory will lead to experiments and so will behave in the scientific method much like an hypothesis. In the end everything speculative has to be decided by experiment, but observation alone does not always determine what has to be introduced into speculation or submitted to experiment.

If there is any merit whatsoever in the account of the scientific method presented by this book, it hangs upon the importance of hypotheses. For the method as outlined here pivots on this concept. There are two stages leading up to the adoption of an hypothesis, and four tests of its validity following its adoption. It is the same hypothesis, crystallized and strongly supported, which still exists in the later stages where it is known as a theory or law. Thus the hypothesis is the focal point of the method. The chief merit of the present treatise is that it shows the indispensability of hypotheses, or the chief demerit if it is not the case.

9. *Adoption*

We have noted already that a productive scientist is one to whom a great many hypotheses occur. He has not only the rigorous training which makes him competent to be a scientist but also the imagination to wander freely among its elements in terms of their possible novel combinations. Obviously, he cannot look into them all. Some appear so silly that they are dismissed forthwith, but enough remain to require a more logical canon of selection. How does it happen that he chooses one over the others to be investigated?

The decision has important consequences, for in terms of the time and energy available to him very few hypotheses could receive such a costly and often extended treatment. Consider for instance what may be involved. The scientist may have to devote most of his remaining life to the task, he may ask the help of co-workers, he may need additional laboratory equipment, he may even need to devise and construct new instruments. If the hypothesis prove false, think of the wasted effort and of his discouragement. This is a chance of course which every investigator takes, but he can cut his losses if he acts wisely and chooses shrewdly at the very outset.

How, then, does he makes his decision? A correct answer is probably not forthcoming. Even the great scientists who have successfully met this challenge cannot always explain how they did it. Yet it is possible to guess. The selection is in all likelihood made in terms of four criteria.

(a) An hypothesis is adopted because it holds out the promise of having a high explanatory value. It has the widest breadth of inclusion of any proposal under consideration, and seems able to recognize under a single rubric the greatest number of disparate facts. That is to say, it seems to have the solid virtue that in addition perhaps to uncovering new facts it helps to explain more efficiently many of the old ones. In this way it suggests the possibility of an enlargement and systematization of the science.

(b) An hypothesis is adopted because it holds out the promise of being consistent with the already established body of knowledge in the science. Consistency with existing laws, a consistency which can be disclosed either by including them or being included by them, is an additional piece of supporting evidence. To suspect such a consistency gives the hypothesis the "feel" of belonging to the science as a legitimate part of it.

(c) Finally, an hypothesis is adopted because it holds out the promise of being conductive to further inquiry. No hypothesis which appeared to be a dead end would be taken seriously. The investigator wants to be able to see by means of the hypothesis how further lines of inquiry would be opened up in case it should prove to be true; in a word, it seems to be a stepping-stone to further exploration. The scientific method is an ongoing process, and whatever facilitates the succession of steps is apt to be chosen over what does not.

A good hypothesis is adopted because it promises to extend inquiry. It has the effect of suggesting lines of inquiry beyond the limits of its own verification. Science is a continuous process, and is not aided by any element which tends to bring inquiry to an end in a given direction. Some hypotheses suggest more than their own immediate verification and lead on to investigations beyond the limits where such verification would narrowly terminate. When William Gilbert set up a function between the magnetic strength of a lodestone and its mass, he introduced a concept, that of mass, which was taken up by physicists who came after him, by Kepler, for instance, Galileo and Newton. Bernard's hypothesis of 1885 concerning the constanty of the internal environment of the organism was succeeded by Cannon's theory of homeostasis with its profitable influence on many aspects of physiology and medicine. Darwin's evolutionary hypothesis gave a tremendous impetus to developmental studies in many biological fields. There is in scientific endeavor a continual search for improvement.

CHAPTER V

THE TESTING OF HYPOTHESES: EXPERIMENT

Hypotheses are tested in several well-defined stages. First, attempts are made to discover evidence for or against an hypothesis by means of experiments. Second, mathematical calculations are made to try to discover evidence for or against the consistency of the hypothesis. Third, efforts are made to discover evidence by means of the predicting and controlling of phenomena, either experimentally or mathematically, or both. In the second stage the hypothesis is regarded as a theory, since it is in a stronger position for having survived the first stage, and in the third stage the theory is regarded as a law, since it is in a correspondingly stronger position still. We shall treat of the first stage in the present chapter and of the others in the next succeeding chapters.

1. *The meaning of "experiment"*

Experimental science is the business of sorting out from all of the propositions related to a particular empirical field those for which some evidence can be shown. A successful experiment which verifies or falsifies an hypothesis may be either a single crucial experiment or a series of experiments in which masses of data are collected and interpreted. In either case, however, what the experimenter has arrived at is nothing but the completion of a stage on the way.

Instrumental techniques to discover factual evidence for or against an hypothesis lie at the heart of the scientific method, since they involve experiments. No hypothesis is acceptable until at least one prediction made from it has been confirmed by experiment. Every general hypothesis admits of an unlimited number of concrete interpretations, and it may be tested by any of these except the ones from which is was derived. An experiment may be defined as a deliberate action undertaken in order to interfere with phenomena in such a way as to compel a decision concerning an hypothesis. By means of experiment a proposed generalization is confronted with the evidence, and this requires observation taking place under controlled conditions, where the investigator anticipates that there will be a certain conclusion reached, quite irrespective of whether it turns out to be positive or negative. Faraday said for instance that he wanted to know how the lines of force were disposed in and about magnets and iron generally, under circumstances of position. This was his hypothesis, and for several pages at the beginning of the sixth volume of his *Diaries* he describes the experiments he performed. Hence an experiment is a refined form of observation, controlled observation, in which some segment of the natural world is forced to an alternative in which it will be as easy to obtain a negative as a positive answer.

Experimental science can be described as planned experience. On the grounds that neither reason alone nor experience along, nor reason and experience together without manipulation, can cope with the intricacies of the natural world, an elaborate logical method has been devised which can weave these three sets of resources together in a fashion which enables them to function at their fullest.

The problem which experiment endeavors to solve can be stated abstractly somewhat as follows.

Given an hypothesis, to deduce the consequences which would be true if the hypothesis were true, and then to design an experiment by means of which the facts can be compared inductively with those consequences.

An experiment, then, is an activity by means of which an hypothesis is confronted with relevant facts. This should not be too strange a situation for an hypothesis: after all it amounts only to a checking of it, one which could already be referred back to its inductive instances, and so it a case of additional sampling, as it were. Experiment does indeed face in three directions: upward toward its hypothesis, backward toward the observed data which suggested the hypothesis, and forward toward the next trial by mathematical verification. The hypothesis itself points the way down to cases and so tells the investigator what he should expect to find. It thus has a predictive aspect. An experiment in this way is a planned action, anticipating certain results which it is expected will be observed. The verification ushers in a series of stages of confirmation and itself constitutes the first of these, thus giving to the hypothesis a first weight.

Experiments are intended to test hypotheses, whether these are implicit or explicit, covert or overt. The hypothesis which guides the experiment may exist for the experimentalist in the form of a reason which has been forgotten, but that does not cause it to exist any the less or to be any the less effective. Herschel has reminded us that it is principles and not phenomena which are the objects of inquiry to the scientist; and this is no less true because the procedure of science requires that the principles be approached through the phenomena. The investigator may be unaware that he is relying upon an hypothesis when he conducts an experiment, but it would be difficult to explain how one instrumental device rather than another, or one problem rather than another, was chosen without it. The very act of selection in the choice of materials to be tested, instruments to be employed in the testing, and conditions favorable to the precise outcome, imply a previous canon of selection in terms of which all the decisions made can exhibit a consistency leading to a certain comprehensive result.

No hypothesis was ever tested in a laboratory but only some of its deducible consequences. And these are tested by attempting to produce specific instances under the consequences considered as universal propositions. For example, if all salts are soluble in water, then sodium chloride can be dissolved. A sample of this salt is accordingly dropped in water and stirred. If it dissolves, it supports the proposition. It can never be determined by experiment whether atomic nuclei disintegrate but only whether some atomic nuclei do, a proposition which stands to the former proposition as particular to universal.

Experiments are methods of ascertaining that the hypothesis which is under examination does not lead to a false result. It is thus as much concerned to discover evidence against the hypothesis as evidence for it. A negative result may be as important as a positive one, as in the case of the Michelson-Morley experiment to determine the drift through the ether. The failures as well as the successes must be scrupulously noted, for they may be equally significant. The selection of instances which are apt to be favorable to the outcome is illegitimate, and certain to defeat the very purpose for which the experiment was performed. Care must be taken, then, to insure that the instances chosen constitute a fair sample, and a fair sample is a random sample.

The degree of verification or falsification obtained must be evaluated, through an interpretation of the data obtained by experiment, which is as difficult a step as the making of the experiment itself. There are experimental speculations as well as experimental procedures and these have to be interpreted: a quantitative assay of a qualitative object, the measurement of an amount of energy, or of the dimensions of a material structure, for instance. The qualitative aspects here are the energy and the material: energy is energy and material is material; these do not change, even though the forms which they may assume and the quantities do. When a given matter is transformed into energy, it is no longer matter. But the amount of substance remains the same. There exist laws which depend upon such irrefrangibility. The laws of the conservation of mass and the conservation of energy recognize the constant nature of entities and processes. Water turns to ice (qualitative change) at 32°F. (quantitative), although the amount of matter is preserved. Results are thus obtained by means of experiments designed to test hypotheses, usually employing instruments and often requiring interpretation. They will usually be of a quantitative as well as of a qualitative nature, where the qualitative aspects are constant and the quantitative as precise as they can be made.

The aim of scientific method is to verify hypotheses. This is often accomplished indirectly, through the failure of the attempts to falsify them. The importance of falsification in this connection is to be found in Aristotle and Galen, it was advanced explicitly by Grosseteste in the thirteenth century and by Popper in the twentieth. Popper's understanding is that the aim of science is to falsify hypotheses, but this is too strong a statement. A falsified hypothesis is discarded, but the aim of science can hardly be said to be to discard hypotheses. Verified hypotheses are continued and tested in other ways and, if found not wanting, erected into scientific laws. If the aim of science were to falsify hypotheses, there would be no scientific laws established. To falsify hypotheses is a negative step taken in the interest of a positive search for hypotheses which can be verified. A strong predisposition in favor of the hypothesis has been evinced in the mere fact that it has been chosen for examination.

The highest peak of objectivity is contained in the attempts of the investigator to find experimental reasons for rejecting the hypothesis which he has been at such pains to establish. An experiment conducted to test an hypothesis but which falsifies it may disclose some very useful information. Not all unsuccessful hypotheses go for nothing. Hypotheses are never absolutely or totally verified by experiment but only partially. On the other hand, a single negative result cannot disprove an hypothesis cast in statistical form but it can prove that what is being tested is no stronger than a probability.

Experiments employ instruments. These are not mere observations at the ordinary common sense level. Either the materials to be tested lie beyond ordinary observational levels, as in the case of refined chemicals which are to be mixed, or the method is beyond that available to common experience, as in the mixture which may be done by an ultracentrifuge. The more advanced the science, the more complex the instruments. The contemporary instruments of scientific research are complex indeed: the electron microscope, the cyclotron, or the digital computer, for instance.

Science is after all a kind of theoretical speculation, and differs from other kinds in having experiment to keep it from going off the track. But it is still theory which leads the way and opens up new prospects to investigation. A completed experiment will include theoretical interpretations; and whereas an hypothesis guides the experiment, a theory guides both. There is a theory for every experiment other than the hypothesis, and it is to the effect that a certain experiment if performed could be an instance of the interpretation of a given hypothesis. The theory of an experiment includes both the hypothesis that the experiment is supposed to test and the relation between the hypothesis and the data which the experiment is intended to produce (as either verifying or falsifying it in some degree). This relation is itself hypothetical and thus constitutes a second hypothesis, which we call the theory of the experiment. When Huyghens in 1678 cut crystals of calc-spar in many different ways to test his hypothesis that light consists in waves, he must have supposed as a second hypothesis that the refraction of the crystal faces would verify (or falsify) his wave hypothesis, and this supposition constituted his second hypothesis and thus served in this case as the experimental theory.

2. *The design of experiments*

In the last chapter we have noted something of the nature and functioning of hypotheses. And we have just been reminded that an experiment is conducted in order to test an hypothesis, and that this is no less true when the hypothesis is merely assumed by the investigator than when it is explicitly stated. An experiment is an action deliberately undertaken in order to answer a question, and if there were no question it is hard to see why there would be such an action.

Which hypotheses, then, are to be tested? The number of possible hypotheses is theoretically limitless; they cannot all be tested, for there would never be enough scientists or equipment or time. On the other hand, any arbitrary restriction upon the discovery of hypotheses is potentially dangerous to the progress of science. We cannot even establish a criterion of verifiability, as some authorities have suggested. If we were to test only those hypotheses that are readily susceptible of experimental verification, we might be overlooking some very important ones. How Einstein's special theory of relativity could be tested was not immediately evident, but that did not prevent its serious consideration and later confirmation. The nearest criterion is perhaps that of conceivable verifiability, but this puts the onus on the imagination of the investigator which will operate as ably without the criterion as with it. What makes an hypothesis a candidate for experimental testing is its suggestiveness, its provocativeness. The same kind of provocativeness which we found in the case of fact is also a feature of some hypotheses, and these are the ones most apt to be examined further.

But some hypotheses are readily tested, while others requires years of hard work. Those which can be quickly tested often are, and are discarded or verified in a relatively short time. However, others require much expensive laboratory equipment and the labor of many men over a comparatively long period before being proved worthy of all the energy expended. Whether or not an atomic bomb could be constructed was in this category. Still others, on which equal effort is expended, may prove inadequate after years of testing. Economy is involved in the selection of hypotheses to be tested. But there are no rules which, followed faithfully, guarantee the best selection. It is in terms of a broad conceptual background that experiments are designed; but more specifically it is the hypothesis which guides the experiment. It tells the investigator what he is to look for and also perhaps something of how to go about getting the results he seeks. Observed facts suggested it, and it returns the investigator to the further observation of fact. Dirac's hypothesis of the existence of positive electrons was made from highly theoretical calculations based on previous observations; the hypothesis was later verified by the discovery of the positron. Hypotheses enable predictions to be made of certain phenomena hitherto not observed; and the observation of the phenomena by means of experiment constitutes verification of falsification of the hypothesis. Bohr's hypothesis of energy levels within the atom – discrete stationary states of quantized energy in which alone electrons could be found – was later confirmed. The hypothesis that particles of matter (electrons) have associated waves like those of particles of light (photons) was suggested by DeBroglie in 1923; he was even able to predict the wavelength of an associated wave for an electron with a given velocity. An experiment is a kind of observation which can tell the investigator something about the hypothesis which suggested it. The only provision must be for the safeguarding of fairness in recording the results.

Once an hypothesis is adopted, it may be said to be on probation. It is ready for testing. It will be examined now in connection with complicated theories and even more complicated facts. It will be studied, examined, subjected to comparisons by means of experiment. The experiment is carried out with a single purpose in view: ascertaining the degree of truth or falsity of the hypothesis. The designing of experiments is a delicate operation requiring the greatest care. The experiment is set up to produce a certain result; without some knowledge of the result sought there could be no design. The instrument chosen and the way in which it is to be used depend upon the conditions which are considered desirable. The combinations and permutations of only a few circumstances would yield such a large number of possibilities that trial- and error methods of choosing experiments would not be feasible for a single problem in the lifetime of a number of investigators. Thus intuitions are necessary in order to get at the essence of the problem. The experiment is more important than the hypothesis now, because the investigator is prepared to give it preferential credence. It is possible to interpret facts wrongly as well as rightly, but impossible to argue with them.

After arriving at a clear conception of the result desired, the investigator has next to state to himself the kind of problem which confronts him in his endeavors to the test the hypothesis experimentally: given the hypothesis, to devise an experiment which will yield results which are relevant to the truth or falsity of the hypothesis. Such a problem calls for concrete interpretation and for solution in detail. For instance, does an instrument already exist, can one in use perhaps be adapted, or does one first need to be invented? Other related subordinate problems are: the selection of a suitable subject-matter, the elimination of extraneous conditions, the choice of a time and place for the running of the experiment, etc. Would this experiment serve the purpose, or would that one do it better? In designing an experiment, it is best to proceed as though a crucial experiment could be set up in every case.

A crucial experiment is one which is adequate to decide the truth or falsity of an hypothesis. In this absolute sense, Duhem is correct in his contention that there are no crucial experiments. But some experiments are more decisive in relation to their hypotheses than others; and it seems legitimate to call the more decisive experiments crucial. Especially is this true in the case of experiments which seem adequate to decide between rival hypotheses.

Experimenting has to do with facts, and this has misled some speculators into supposing that the scientific method is confined to facts. Chief among these is Francis Bacon, who believed that deduction plays no role in the scientific method. Bacon thought that because science begins with facts, it must be held down to facts. He did not understand that starting with and being held accountable to facts does not mean remaining with facts nor preclude framing a description of facts in terms of abstractions. To confine attention to the facts, on the assumption that if a sufficient quantity of them is collected it will yield its own hypothesis, neglects the criterion by which they were gathered in the first place. This involves a fallacy and has been named "inverse probability."

The experimental verification of an hypothesis usually develops with painful slowness. If an experiment is successful, it will be repeated by the investigator before it is published and by others after the publication appears. Scientists come only gradually to trust an hypothesis, but when they do they refer to it as a theory. The more radical the hypothesis, the more difficult its acceptance; for in science all depends upon evidence and nothing upon convention. No law is so securely established that nothing could shake it; all must yield before the testimony of the facts.

An experiment is a comparison; the investigator endeavors to match the proposition which the conclusion of the experiment constitutes with the proposition which is the hypothesis. If he detects a similarity, that is affirmative evidence, strong or weak; if he detects a contrast, that is negative evidence. For an experiment is a proposition exemplified and dramatized. This is the direct method, but there are probably an equal number of important instances where the method has to be indirect. If the proposition on probation, which the hypothesis is, does not lend itself readily to test by experiment, deductions from it might do so. The hypothesis of an organic chemist, for instance, may be the structure of a new substance which he has succeeded in synthesizing. Such a structure if his guess is correct would have certain chemical properties which could be verified by testing – by qualitative analysis, determination of molecular weight, quantitative analysis, the establishment of atomic linkages, etc. The Schrödinger wave-equation has never been directly tested, but a sufficient number of mathematical deductions from it have been made – enough, that is, to justify considering it verified.

It often happens that very ingenious methods have to be devised to test a hypothesis. Elaborate instruments have to be designed and built, or, if the experiment is in the field, sometimes travel to distant places is necessary to obtain the conditions proper for observation with instruments. In 1969 special instruments were designed to detect the weak gravitational waves that have long been suspected to exist, and it was thought necessary to place these instruments 600 miles apart. A further hope in connection with the same experiment is that one such instrument can be located on the surface of the moon.

The discovery and verification of hypotheses is the main line in scientific research. There are other types of experiments, however, those of a rudimentary nature and those incidental to the main line, for example. Experiments may be conducted merely to determine existing conditions; an experiment is designed in the expectation that something not known will be discovered. Harvey's experiments on the circulation of the blood were of this sort, although he no doubt had the analogy of the pump in mind. Pumps were fashionable in England and Holland at the time, and irrigation was already a familiar procedure in agriculture. Some experiments are undertaken in the hope that they will suggest hypotheses. This is the so-called trial-and-error method. When an experimental subject in biology which could reproduce in a short period and therefore provide a sufficient number of generations to test genetic material was wanted, Castle in 1906 came upon the fruit fly, *Drosophila.*

Experiments are also conducted very often merely to determine whether conclusions arrived at by other investigators are as reported. No crucial experiments are allowed to stand alone; all are repeated by subsequent workers, who often produce refinements in instruments, in techniques, and even in the accuracy of conclusions.

3. *The logic of experiments*

There is a pre-logical stage which consists in the preparation for such procedures, and this is the stage of recognition. Mere observation, as we have already noted in a previous chapter, involves the fitting of a case under a class, the recognition that an individual is a member. Recognition takes no logical step but does prepare the observed material for logical treatment. In 1932, a nuclear particle of a new type was discovered in the same way that the proton had been discovered: it was knocked out of the nuclei of atoms by bombarding them with alpha particles. It carried no electric charge; and so it was named neutron. The investigator had isolated and tagged a type of actual object, the neutron, a constituent of the nucleus. He was now prepared to study its other properties. In this way facts are prepared for relation to propositions.

The verification (or falsification) of an hypothesis by an experiment may be arrived at deductively from the hypothesis or inductively from the experiment. Deductively, it is argued that if the hypothesis is true, such-and-such ought to be the case, with an experiment subsequently devised to determine whether it is the case or not. Inductively, experiments are performed, and if successful are shown to support that hypothesis which is similar in character but more general.

An experiment is a particular proposition translated into action. The translation is never complete, for particular propositions contain a generality never possible to action, and action contains a quality never possible to propositions. The incurable generality of language, which characterizes even the most particular propositions, even, say, the most concrete statements describing laboratory experiments, insures that hypotheses, being universal propositions shall never be absolutely and conclusively verified or falsified. We say that a proposition is empirical if it can be referred to actual fact by means of action for the determination of its degree of truth or falsity. Since the equations of physics and chemistry are always absolutely stated as universal propositions and the descriptions of experiments are reported as particular events uniquely dated and located, the theories of science can never be exact descriptions governing the data disclosed by experience but can only state the ideal conditions to which those data indicate approximations. In other words, we learn from our experiences according to the scientific method something of the conditions under which those experiences take place.

An experiment, then, can be considered in its role as an activity or as a logical relation. As an activity it is essential act, a something happening that is also a doing. As a logical relation it is an act which when translated into propositional language verifies or falsifies an hypothesis. As an activity, it is an event, and from the standpoint of its relation as a whole to its own parts not dependent for its meaning upon other events. But as verification or falsification it is related and so has a meaning; the event is related to other events in a way which affects them as regards their meaning. In the verification stage of an hypothesis, the hypothesis is so to speak brought back down to the earth from which it first arose as a proposition when its generality raised it above the singular which first suggested it as a possibility. But now it is no longer the same kind of singular, for the new singular, which is the experiment, it not a flat one but looks upward to further stages, and can be expected (if verified rather than falsified) to have its status of generality reinforced.

Logically, an hypothesis might be considered the major premise of a syllogism and an experiment the minor premise, with the conclusion the result of the colligation. The experiment is intended to bring the universality of the hypothesis to the test by relating it to individual fact. The logical process of colligation is essential to the drawing of a conclusion from these two otherwise disparate elements. If successful under this scrutiny, hypotheses are partially established and come to be regarded as theories. However, theories can never be absolutely anchored in facts but only relatively so, because the facts can only be sampled and never exhausted.

Herschel, the astronomer, saw clearly that the scientific method demands continually the alternate use of both the inductive and deductive methods. In the scientific method the relations of facts to propositions may be of two varieties: they may give rise to propositions by means of induction, or they may be deduced from propositions by means of deduction. The first step occurs in the discovery of hypotheses and the second in the testing of hypotheses by experiment. But there is another kind of induction in relating facts to propositions when an investigator argues from facts already known to an hypothesis under consideration. We shall call the first kind of induction the induction of hypotheses and the second kind the induction to the verification of hypotheses. The first kind is exploratory, and the second kind confirmatory. The first kind is without regard to probability; the second relies entirely on probability. It makes no difference to the brilliance, the sweep, or the incisiveness of a scientific hypothesis whether its discovery by means of induction was from one instance or from a million: the hypothesis will be the same in both cases. But to the kind of induction which consists in the verification of an hypothesis, it makes a very great difference whether one argues from one instance or more; and just how many is significant, for in seeking support for an hypothesis the number of favorable instances examined and the proportion of favorable instances may be extremely important.

Some investigators have identified induction with probability, supposing that all inductive reasoning takes place in terms of probability. But this would be true only if all inductions were of the second kind described above, namely, induction in verification of hypotheses. But there are other types, as we have noted; in particular, there is induction of the first type, the kind of induction that occurs when hypotheses are first discovered. When a single instance suffices to suggest an induction, probability hardly applies. The inductive step testing a hypothesis requires beginning with the adoption of the hypothesis, and then, by means of the predictions which it usually makes, trying the experiment. The actual operation relies upon the relation that if the hypothesis is true, bringing about the existence of certain conditions ought to have certain results. The experimenter then brings about the conditions and notes the results. If they are favorable, they verify the hypothesis to that extent. It is not difficult to think of recent illustrations of inductions made from single instances. Planck's quantum constant occurred to him as the result of a single observation. Einstein's special theory of relativity came to him as the result of reasoning from a collection of disparate and isolated phenomena which were not at all statistical. Crucial experiments are usually isolated and individual, and not matters of averages.

Although an experiment is a single event, it is not considered unique but rather the representative of a class of similar events. The investigator is not dealing with single events for their own sake but through them with a general class of events: experimenting is always sampling. Given the proper equipment and the proper procedures, the same results will be obtained, this is what is presupposed behind the performance of an experiment: and when the experiment is one of sufficient significance the scientist is careful to see that it is repeated a sufficient number of times to support the assumption concerning the class.

There is a rigid rule to the effect that the same data were used for suggesting the hypothesis may not be used in testing it. The resistance of a single beaker of fused quartz to rapid temperature changes suggests that all chemical ware should be made of fused quartz. But this same beaker cannot be used in confirmation of this hypothesis, instead flasks, tubes and dishes will have to be sampled. Otherwise the argument would be circular. If several hamsters whose skin has been shaved and rubbed with a solution of nicotine were to develop carcinoma, then the hypothesis that nicotine is a carcinoma-producing substance, will have to be tested on random samples of other populations.

An experiment may mean logically that an hypothesis is being inductively discovered, or it may mean that an hypothesis is being tested both inductively and deductively. There is no mechanical method of calculation for the deductive step from the hypothesis to the experiment that would test it for verification. In 1935 H. Yukawa in Japan predicted the discovery of mesons, and one variety – μ mesons – was recognized in cosmic radiation three years later by Anderson, Neddermeyer and others, and another variety, π mesons, was recognized in 1947 by C. M. G. Lattes, and produced artifically in the laboratory of E. Gardner and Lattes a year after that. The prediction of mesons was a deductive step, yet considerable ingenuity was required to detect and then to generate the mesons to further support the verification of the deduction.

Before experiments can test hypotheses, they must have been deduced from hypotheses. If an experiment is successful, it will furnish inductive support for its hypothesis. The deductive step, which is taken first, is from the hypothesis to the relevant data; the inductive step, which is taken second, is from the relevant data to the hypothesis. Thus experiments are both inductive and deductive.

There is no such thing as the absolute proof or disproof of an hypothesis by experiment. If an experiment is successful, then the hypothesis is verified to that extent, and to what extent depends upon the nature of both the hypothesis and the experiment, for in some cases it will be more crucial than others. We do not speak of experimental proof but only of experimental support: it is always a matter of degree. But in no case will the evidence from a successful experiment be considered sufficient to declare the experiment proved. If the experiment should be unsuccessful, then we do not know to what extent the hypothesis has been falsified, only that there has been some evidence in favor of its falsification.

Ordinarily, in logic a negative particular proposition contradicts a universal affirmative proposition, but when the particular negative proposition is represented by an experiment this is not always exactly the case. Logic may be translated into action and action into logic, but the action more often than not lacks logical rigor and in the actual world in which experiments take place there usually are extenuating circumstances. Thus the interpretation of an experiment may show that it calls for the abandonment of the hypothesis but only for a revision; and it has sometimes happened that an hypothesis as stated turned out to be not false but inadequate, and that a reformulation may save it. The decision in every case rests upon the evaluation of the data in that case, and thus far we do not have any established standards for making such judgments but rely upon inclinations acquired from repeated experiences of a similar sort.

The value of an experiment to the method of inquiry lies in its crucial nature as often as in its verification or falsification. Falsifying experiments have at times been suggestive, such as the Michelson-Morley experiment to detect the drift of the earth through the ether. In logic, falsity can lead to truth (but not truth to falsity). In the applied logic which is the scientific method, this also holds. If it had not been possible to set up such false hypotheses as phlogiston, caloric fluid, and the ether, something closer to the truth in those areas may never have been discovered. Knowledge is not preceded by ignorance usually but rather by false knowledge. Again, it often happens that the results of successful experiments have seemed insignificant, where the design of the experiment was not such as to produce a significant outcome in any case. Many experiments in the social field, which include the careful gathering of enormous quantities of statistics, are of this latter sort, for there is little conclusion either of a theoretical or a practical nature.

In the scientific method we are often given the experimental data and asked to find the explanatory principle which will account for them. But sometimes also we are given an explanatory principle presented as an hypothesis and asked to find by means of experiment the data which will support it. Hypotheses may be supported, revised or abandoned, in accordance with newly discovered data, but it is also true that facts which are well known may have to be reconsidered in accordance with newly discovered hypotheses, that is to say, the facts may be selected, classified or re-interpreted in the light of a new insight into their nature presented by the hypothesis considered as an explanatory principle. Thus there is constant interaction – data call for revised hypotheses and hypotheses call for reinterpreted data, always involving the process of experiment and linking by means of experiment the individual data and the general hypotheses.

Mill undertook to show by means of his canons of agreement and difference how the logic of experiments worked as a mechanical method of discovery. But he succeeded in showing only a negative and not a positive value; his canons give the necessary but not the sufficient reasons for regarding a connection as reliable. The evidence considered sufficient to prove cause only indicates that causal factors may be present and should be looked for. Moreover, they deal with occurrences regarded as revelatory of causes and ignore all types of relationships more complex than concurrence or its absence. Functions and correlations of more complex sorts, are often so often involved that Mill's method may not be considered guides to good laboratory procedure except in the narrowest cases.

Mill's canons purport to be dependable formulas for the discovery of empirical cause and effect. No phenomenon can be regarded as a cause if the effect is present in its absence or absent in its presence or if when it varies the effect does not vary in a corresponding manner. They tell the experimentalist what he will find to be true provided a cause-and-effect relationship exists in some phenomena under investigation; but they will not tell him how to find the cause. They are thus not methods of inquiry but rules of inspection for the results of inquiries once they are made. They have the negative value of eliminating faulty hypotheses by means of experiment, and are thus worth knowing, especially in the early stages of the scientific method.

A more advanced type of thinking which has been useful in applications to the scientific method is the mathematical function. A function describes how one quantity, y, depends upon another, x, and so involves the use of variables. In the scientific method, and particularly in experiments, there is a constant search conducted to discover independent variables in order to account for the behavior of dependent variables. The experimenter is the one who attempts to guide the independent variable over typical samples of its full range, for then limiting cases may be uncovered which indicate the fruitlessness of looking in this direction. Most time and motion studies for example can best be understood in terms of mathematical functions.

4. *Experimental criteria*

Scientific experimentation means interference with the subject-matter under investigation in order to bring about a certain kind of observation. The procedure involves conjectures designed to avoid contradiction with fact, and involves the satisfaction or a certain set of conditions. A good experiment must be (a) isolated, (b) analytic, (c) repeatable, (d) crucial, and (e) heuristic.

(a) Experiments should be properly isolated. The necessary degree of isolation is in fact a prime condition for the conducting of an experiment. The investigator must be as sure as possible that all uncontrolled variables have been eliminated or randomized. These are often innocent-looking factors which are capable only too often of nullifying the results of the experiment. Atmospheric conditions such as temperature, atmospheric pressure, dust, also stability of the laboratory containing the instrument, the amount of available light, these are some of the factors, and there are others; for instance, minute quantities of contaminating material, unwanted variations in biological specimens, unrecognized instrumental defects. The elements which may be possible sources of interference vary, and what may be trivial in one experiment could prove disastrous in another. The incident light is a negligible factor in a learning experiment conducted with mice, but could be fatal in an interference experiment conducted with photons. With some chemical experiments the investigator might very well neglect all consideration of the time of day, but with many biological and psychological experiments he could not do so.

All science is a matter of approximations. The degree of caution which has to be exercised in ascertaining that all undesirable factors have been eliminated from an experiment, that small or large deviations which might cancel or negate the results, have been sufficiently ruled out, can hardly be overemphasized. Ideal conditions are of course unobtainable, but we can approach as closely as possible to them, and this asymptotic approach to the ideal requires endless patience and constant vigil. Astronomers always seek to make their observations when the object under scrutiny is at the meridian. Also, they seek to obtain conditions for observation as free as possible from interference. Observing instruments have been moved from sea level to mountain top to balloon, and now moving them to artificial satellites is being attempted. There can be no relaxing the watchfulness of the investigator who expects to obtain from his laboratory a relevant and meaningful result.

(b) Experiments should be analytic. That is to say, they ought to take place in such a way as to reveal properties beyond ordinary observational levels. Human beings stand between the microcosm and the macrocosm, between one world which is too small and another which is too large to be reached by ordinary methods of sensation, action and thought in common experience. Both worlds are immensely rich, varied and complex, and possibly without end. The investigator from his mesocosmic perspective looks in both directions by means of techniques of analysis named the scientific method. Science takes off from common experience, but the results may be very startling to common experience, thanks to the use of instruments and mathematics.

We shall have much more to say on these topics later on in the chapter. From Galileo's crude early telescope to the most complex of modern synchrotons, the instrument employed in experiments is a guarantee that the results of the experiments shall be nothing that could have been learned in other ways at the level available to the unaided senses. And from the early formulas of Euclidean geometry to the most complex formulas of partial differential equations and the tensor calculus, the mathematics employed in experiments guarantees nothing that could have been learned in other ways at the level available to the unaided senses. For it is the knowledge of mechanisms that science is seeking and not the mere recording of effects. We see the effects all around us without special effort. But the mechanisms lie deeper or farther off, and it is to probe them that the scientific method was first devised.

(c) Experiments should be repeatable, and most important experiments have been repeated. If an experiment is a way of compelling some segment of the natural world to answer a question put to it, then we have reasons to suppose that the same answer will always be given to the same question. The replication of experiments is as important as the experiments themselves. One instance of an experiment can never be taken as indicative of anything except perhaps a signal to repeat the experiment. The value of experiments – it is almost possible to say the entire value – depends upon their repeatability. What is learned from an experiment would be worthless were it not for the assurance that what happened once can be corroborated any number of times. And this is not left to faith but is confirmed in each instance: no matter how carefully an experiment may have been designed and executed, if an important hypothesis depends upon it, it will be repeated, and repeated not only by the same investigators but also by others working with similar material and with instruments made to the same pattern.

As a matter of fact, this is the established practice in scientific procedure. The repetition serves to ensure the elimination of experimental error and often produces refinements of observation that might have gone undetected in the original trials. Michelson's experiment in 1881 to detect the ether drift was repeated by him in collaboration with Morley in 1887. The experiment showed that the velocity of light is independent of any movement of its source, and further that the uniform motion of a material body cannot be detected by observation of that body alone. It was again repeated by Morley and Miller in 1905, in 1926 by Kennedy and in 1930 by Joos.

The repeatability of scientific experiment is responsible for what has been called the self-corrective nature of the scientific method. What is proposed by an hypothesis cannot be established by a single experiment; the scientific method makes this impossible because of its demand that all experiments shall hold themselves repeatable and that crucial experiments must be repeated. It is perhaps by this criterion alone that science distinguishes itself from comparable enterprises. Every work of art is unique, and so is every religious revelation; but science decries uniqueness in this regard, and renders every experiment available to every investigator, insisting that for any findings which are held even temporarily true (and none are held more than that), the same results must be obtainable at any date and place.

(d) If possible, experiments should be crucial. Ordinarily a crucial experiment is understood to be one which by itself is able to render a verdict in the case of an hypothesis. The phlogiston theory of Stahl was accepted for almost a hundred years until the devastating weighing experiments of Lavoisier, which showed that although matter may alter its state, it does not alter its weight, was directed against the theory. As Popper correctly asserts, a crucial experiment is apt to be a falsifying experiment. An experiment which gives an affirmative answer usually requires further support. A crucial experiment can also be one which decides between alternative hypotheses. But there is also the experiment which uncovers a significant fact and is crucial in this way.

It is not the purpose of experiments to bring out just any facts. An experiment ought to be conducted with a precise idea of what is involved. What the scientific investigator is after are provocative facts, those which have an importance to research beyond the narrow meaning of the single datum. The range of the degree of support (or of rejection) furnished to an hypothesis by an experiment is a very wide one. One more conditioning experiment can now do very little to reinforce a learning theory which is already fairly secure. But if any investigator were able to devise an experiment which could verify (or effectively falsify) Einstein's unified field theory, that would be very welcome indeed. It is necessary to remember at all times that it is important to save the phenomena but not important to save the hypothesis. An experiment that comes at an hypothesis in an unique way is more apt to be crucial than one which simply repeats with small variations an experiment which has already been checked a number of times. Thus ingenuity in the devising of experiments is just as much of a contribution to the scientific method as originality in the discovery of hypotheses.

(e) Experiments should be heuristic; they should open up avenues of research and pose more questions than they answer. The invention of a new experimental instrument or technique will have this effect. The invention in 1800 by the Italian, Volta, of the primitive battery, which consisted in a pile of discs of zinc, copper, and paper soaked in brine, suggested the electrolysis of water in the same year to Nicholson and Carlyle in England, and so led to the entire science of electro-chemistry. Von Laue's use of a crystal as a grating for the diffraction of X-rays, and the 1927 experiment by Davisson and Germer who shot a stream of electrons against a nickel crystal and by thus producing waves began the science of wave mechanics, are instances.

An experiment may be trivial – or it may begin an entirely new scientific development. Freud's treatment of a patient with conversion hysteria constituted an experiment which suggested that the hypothesis upon which it rested could be explored much further by means of the same technique, and so started the theory and practice of psychoanalysis. The scientific technique calls for skill in discerning the significance of a single datum. The observation that the surface of a liquid in a narrow tube gives a concave meniscus led to the understanding of the forces of interaction between the atoms forming the solid matter of the tube and the atoms forming the liquid, and so to the capillary theory of Laplace. In studies of black body radiation, the attempt to account for the equilibrium distribution of electromagnetic radiation in a hollow cavity led Planck to assume that the energy of an oscillator of frequency ν is restricted to multiples of a basic integral $h\nu$. The energy of the oscillator is then $E = nh\nu$, where n is any integral from 0 to ∞.

5. *The use of instruments*

In experimentation with instruments all three ways of controlling observation are employed, as we have noted in Chapter II. The object is adjusted in order to bring it into focus, the ground is prepared for observation in such a way as to reduce to a minimum the degree of interference, and the conditions for observation are selected. The controlled character of instrumental observation is its most obvious feature. Controlling observations means first of all inventing instruments and this in turn depends upon the skills available to make the instruments and to devise the techniques for manipulating the materials. Thus science at this stage is reliant upon technology. In the mode of discovery, technology serves pure science. We shall see in a later chapter that in the mode of application, pure science serves technology.

Generally speaking, instruments translate microcosmic and macrocosmic events into events in the mesocosm with which our limited human senses can deal effectively. A good example of the translation of microcosmic events is the fluorescent screen which presents invisible ultraviolet radiation as visible wave-lengths. A good example in the case of macrocosmic events is the use of infra-red filters with optical telescopes set for time exposures which produce photographic plates recording astronomical objects and interactions not otherwise available to sight. Instruments make possible certain observations which would be impossible without them. By means of an extremely sensitive and ingenious electrical balance which required three years to design and construct, R. A. Millikan in 1911 was able to determine the mass of an electron. Gases could not have been analyzed without the mass spectrometer, radiation could not have been detected so easily without the Geiger-Muller counter, and changes in electronic potentials could not have been seen without the cathode-ray oscillograph. It is a long way from the instruments available to the eighteenth century chemists: the crucible, the flask, the retort and the balance, to those which the modern chemist now uses: the absorption spectrophotometer, the electronic analytic balance, the refrigerated ultracentrifuge, and the photoelectric colorimeter. But for the problems the early chemist worked on his instruments were adequate. For instance he wanted to know the relations of gasses to solids, to metals for instance, and he wanted to be able to distinguish between gasses and determine their properties.

Observation as exercised at the stage of experiment must be carefully distinguished from observation as the search for data, as outlined in Chapter II, above. Admittedly, there are similarities as well as differences, and so perhaps it will be helpful to mention the similarities first. At both stages complex instrumentation may be employed. It is useful to remember also that observations may be made at any and every stage in the scientific method; science never allows itself to become very separated from the phenomena. The search for data is the starting-stage in the process leading to the *discovery* of hypotheses; experiment is the earliest stage in the *testing* of hypotheses.

But despite these many similarities there are recognizable differences between the complex observations conducted on the results of complex experiments and the simpler observations which are made in the search for data. The observations of experiments are planned with the experiments; the operator knows before he starts what sort of data he will be looking for. Then, too, many of his results occur in the form of pointer-readings on the dials of instruments, which he must then interpret in terms of what he knows he is testing. The observation of instruments is quite different in character from the observation of natural occurrences in which there has been no human interference.

Advances in science are of many kinds, but chiefly of two: advances in theory and advances in instrumental design. And these are not unrelated. It often happens that in order to conduct a certain experiment the choice of instrument may first have to be determined. A whole set of preliminary explorations, of alternative hypotheses, of designed and performed experiments, and of the calculations of results, might be necessary before the experiment in view can be tried. If the instrument does not exist, then it will have to be invented and manufactured. This has been the case with explorations on ocean bottoms, where the bathysphere was the instrument needed, and with the explorations in outer space, where multi-stage rockets were required. It often happens that an instrument does exist but advances in design greatly facilitate the observations. The zonal rotor designed by Anderson improved the technique of centrifugation and made possible the separation of subcellular components in much greater quantities than had been possible before with swinging bucket rotors.

It would be difficult to overemphasize how frequently and intimately observation serves to tie together theory and instrument in the scientific procedure. Not only does the investigator make the instrumental observations necessary to suggest hypotheses, but when he is testing the hypotheses by means of experiments with instruments he observes again; for both are necessary to experiments. The apparent recession of the nebulae, according to which the more distant the stars the faster they are rushing away, began as the hypothesis of recession based on observation, first of the nearest Cepheid variables, next in terms of the luminosity of star clusters containing Cepheid variables, and finally just of the luminosity and size of the nebulae themselves. But the theoretical hypothesis that the distances to nebulae are functions of their radial velocities arose from observing that the lines of the spectra move toward the red end of the spectrum: the farther away the nebulae the faster they recede. The data supporting the theory of the "red shift" was due to the technique of photographing the results obtained by combining the spectrum with the reflecting telescope. And now further confirmation has come with the radio telescope.

The use of instruments is largely in the service of observation, but there are two kinds of instrumental observation: direct and indirect. Direct observation involves the use of instruments to extend the ordinary human senses of the investigator: the microscope or the optical telescope, for instance. Indirect observation is where the instrument substitutes for the senses; *it* does the observing, and the investigator then observes the instrument: the spectropolarimeter, the cathode ray oscillograph or the liquid scintillation counter. Thus there is observation both by means of instruments and of the instruments themselves.

Direct observation by means of instruments carries the observer into new dimensions of the external world, and makes available to him things and events smaller than his unaided senses could detect or more distant than they could reach. The electron microscope and the reflecting telescope bring within his focus both inaccessibly small and unattainably remote objects. Elusive processes and otherwise undetectable forces can be sampled and measured with the help of instruments. Effects at ordinary levels are better understood when instruments disclose the presence of causes at more analytical levels. What had been thought to be heuristic entities may prove to be empirical and even observable when examined under high magnification, as was the case when genes were observed in chromosome chains by means of photographs taken with the electron microscope.

Indirect observation takes place in a number of ways. The most familiar perhaps is that in which certain effects are recorded on an instrument, and then presented to the investigator in the form of pointer-readings. Another kind is the direct observation of the effects of phenomena, which are in this fashion indirectly observed. The spectra of light rays, which under certain conditions of thermal or electrical disturbance are emitted by the atoms of elements, are characteristic of such atoms and so tell us something about their structure. No one can see the alpha particles which pass through the Wilson cloud chambers, but the condensation nuclei formed by the positive ions are visible in the water vapor. Time exposures made through optical telescopes with the aid of special equipment as infra-red and ultra-violet filters are indirect observations, for although the photographic plates are directly observed, what they represent could not have been observed at all with the unaided senses. The tracks of alpha particles in the Wilson cloud chambers and the photographs taken with the two hundred inch Hale telescope at Mount Palomar are equally good examples of indirect observation.

Instruments assist observation; but they also measure and manipulate materials. The Geiger counter registers the number of individual particles, photons, for instance, which it encounters, and the ultra-centrifuge divides particles in suspension. Measuring instruments are indeed plentiful. The photometer to measure the luminous intensity of light, the galvanometer to measure small electric currents, the calorimeter to determine quantities of heat, the hemocytometer to count blood cells, are common examples. But manipulating instruments are no less common, for instance the larger accelerating devices which hurl subatomic particles with enormous speeds at the atomic nucleus.

A further degree of penetration is accomplished by combining instruments. Cameras and light filters of various kinds are attached to telescopes, and they may be attached in much the same way to microscopes. Electronic computers are hooked up to devices for tracing nuclear particles (cloud chambers, bubble chambers, etc.). In neurophysiology it has been possible to combine the problem-box of the psychologists with the electroencephalogram of the physicists with the conditioned reflex techniques to learn something about the cortex. Indirect observation becomes the most important sort of observation as a science advances into areas not directly observable. Cyclotrons, oscilloscopes, Geiger counters, supplement the more direct work accomplished with the photographic plate, the scintillation screen, the ionization chamber.

It is easy to understand that no instrument was ever designed without a use in mind, and the use to which an instrument is to be put is decided by the necessities of testing hypotheses. It may happen, of course, that an instrument designed for one purpose proves useful in another. The interferometer designed by Michelson and Morley for the purpose of measuring the motion of the earth through the hypothetical ether has also been used in astronomy to break up the light from eclipsing binaries. The interferometer doubles the power of the telescope and so makes it possible to measure the light of stars twice as close together as the telescope alone can resolve. Again, antennae designed for earth-bound communication has been turned skyward and adapted to the purposes of the radio telescope, thus opening up an enormously fruitful new avenue of exploration.

No one knows when an instrument is invented where it will lead, for each has possibilities of its own. A new instrument is a new opportunity in science, for built into every one there is an entire host of silent hypotheses awaiting trial. Thus science advances on a common front, alike whether it has discovered a fresh conception, a novel technique or a unique instrument; for conceptions have to be tried, and this cannot happen unless a technique informs the investigator about the use of an instrument.

It often happens that instruments are devised which have been called for by no specific hypothesis. Can it be said that such instruments did not arise in response to the necessity for testing an hypothesis? Hypotheses come in various shapes and sizes. The development of some instruments must await certain theoretical and technical developments to make them possible at all. Metals exhibiting resistance to high temperatures can now be tested, but this was not possible before. The electron microscope with its magnetic lenses could not have been constructed until electrons and protons with their associated waves were shown to have the capacity to diffract light just like X-rays. But after it was built and improved and put to many uses, one could ask, in response to what hypothesis was the electron microscope first constructed? And the answer would have to be as follows. *There are objects at analytic levels below the range of the optical microscope.*

Without such an hypothesis (which was no less present for having been overtly expressed), no such instrument would ever have been put together.

The design of instruments begins with the specifications stage: a definite description of the materials to be used and of the form in which they are to be constructed. The properties of a good instrument must include at least the following: high speed response, making possible fast counting and timing; wide-range recording; low level of interference, for instance a minimum of "noise"; a full grade of resolution; fineness of discrimination; ease of manufacture, as provided for by simplicity of construction; and adaptibility, which is to say, great range of application to different sorts of problems and subject-matters.

Often it happens that new advances in instrumentation are made possible by the discovery either of new materials or of new properties of old materials. A new principle also may be involved; though more valuable, such an ingredient is also rare. The addition of electronic detectors to scintillation counters improved their efficiency millions of times, so that now this instrument, which lay neglected for several decades after its invention by Crookes in 1903 and Regener in 1908, begins to threaten the popularity of the Geiger counter. A number of factors may be involved which bear on the outcome of experiments made with the instrument and which therefore affect its design, as for instance its imperviousness to the elements which are to be used in connection with it. Otherwise, the instrument itself will disturb the requisite isolation. Chemicals which corrode containers in which they are to be mixed or separated, apparatus with frictional or magnetic properties which could interfere with the conditions established for an experiment, are examples. An apparatus which is unwieldy, such as the earliest digital computers, or one which lacks the necessary stability, contributes to the difficulties.

A great degree of precision must be built into every experimental instrument. Calibration is a crucial step in its preparation, and this must be carried out in conformity with acceptable standards of reference. The perfection of techniques themselves may influence the standards, as occurred when the yellow lines in the cadmium spectrum were substituted for the gold bar maintained at a constant temperature in Paris as the reference for meter length. Further and more refined versions of the same instrument may be developed, as for instance in the case of the goniometer to measure the angles of crystals, when in place of the primitive contact goniometer there was substituted the more advanced reflection goniometer. The performance of microscopes and telescopes is continually being improved with respect to their magnification, resolution and contrast.

A word of caution should be interpolated here. Generally speaking, the instrument employed need not exceed the requirements of the problem under study. In an experiment in physics involving the use of a diffraction grating, where one containing a few hundred parallel lines to the centimeter will do, there is no need to employ one containing thousands, even though such fine gratings do exist. There is no need to build a sweep vernier into a machine when the calculations will not call for a subdivision of the main scale. Technological ingenuity is a necessary part of the experimentalist's equipment as well as the intuitive perception of the degree of accuracy required in order to make a particular experiment a success. However, there is no need to exercise precision beyond the degree of fineness of the experiment. The electron microscope is not called for in making blood tests nor Riemannian geometry in making terrestrial measurements.

A similar point is contained in the warning against the superfluous use of complicated apparatus. Science was made possible by the balance between the use of instruments and the mathematical interpretation of the results so obtained. But there is the opposite extreme which consists in supposing that because instruments are indispensable, the better the instrument the more scientific the results. There is no need to employ a cyclotron such as the big one at Berkeley, when a Van der Graaff generator will do the job just as well. For certain astronomical observations no more is accomplished by the two hundred inch Hale telescope than would be done by an ordinary twenty inch reflector.

Instruments designed for prolonged operations in protracted experiments ought if possible to be automatic. The use of control mechanisms, servo-mechanisms, and feed-back arrangements such as thermostats, barostats, electronic voltage regulators, are desirable in this connection. Part of the automatic setup is recording equipment. The use of such instruments as pen and paper, plastic discs, magnetic wires or tape, scratch marks on coated glass or celluloid, which can be read off and also filed away for permanent reference, is highly valuable. The ideal instrument would set itself in motion, maintain its operation automatically with provisions for corrections in the case of errors, and make a permanent record of the results. It is possible for machines to maintain experimental schedules which would be quite impossible for human beings, as with clock-driven telescopes and electronic computers. Then again, instruments can often go to places inaccessible to human operators, such as when recording instruments were carried aloft by artificial satellites fired in 1958. The direction of science away from the fallibility of the human senses, as in the case of the substitution of the thermometer for the skin and of the vibrating membrane for the ear, which occurs in the early stages of science, are extended in later stages by the substitution of automatic experimental mechanisms for the entire human experimenter.

The addition to instruments of apparatus intended to bring the results up to detectable levels also occurs. Amplification by means of levers, mirrors and vacuum-tube amplifiers, the magnification of sounds and current electronically, are instances of this method. At the other end, however, there is the limitation which exists for all instruments in terms of general interference and failure – noise. It is impossible to eliminate, though often possible to reduce, the element of interference caused by features of the instrument itself. Chance distortions occasioned by extraneous material, spurious and undesirable signals interspersed among the others, errors increasing the degree of uncertanity, are always present in some amount, and mark by their presence the limits of effectibility of the instrument.

Particular types of instruments have their own inherent limitations. The cloud chamber makes clearly visible the tracks of alpha and beta rays but will not record gamma rays. That which makes it possible for an apparatus to function also provides it with limits to that function. Mitosis, the process by which the nucleus of a cell splits in two, cannot be observed by means of the electron microscope, because the beam of electrons has to be focused in a vacuum, in which living material cannot survive. To increase the magnifying power of a microscope or telescope, say, is to decrease the brightness of the image; so that there is a point past which the usefulness of the instrument is impeded. The only recourse in such a case is to turn to some other type of instrument. The electron microscope can do what the optical microscope could not; and the radio telescope supplements the reflecting telescope.

Many instruments are actually measuring devices, all those, certainly, whose results have to be read off a dial. Pointer readings are measurements, and almost every instrument has an assigned metric. The most obvious case of a measuring instrument is the computer. Analog computers are used in medical research, nuclear reactor design, meteorology, chemical kinetics, and in a number of other sciences.

6. *Measurement*

The two devices essential to most scientific experiments are: instruments and mathematics. Many of the mathematical instruments were designed to take measurements. The use of mathematics in the scientific method however is confined neither to instruments nor to experiments. It is to be found most commonly in the form of measurement. The measurement to be discussed now is not the pure measure theory of mathematics but rather applied measurement as employed in the empirical sciences. For it is in the discernment of measurement that scientific inquiry reaches that degree of combination of the general with the specific (the interpretation of the results of experiments by mathematical means) which enables it to discover laws.

Measurement is comparison with a standard; more specifically, the comparison of some magnitude of matter or energy with the magnitude of a standard chosen for the purpose. The standard is established in convenient units, and the comparison is made between the standard and an object or event. The investigator singles out certain properties of objects but needs to manipulate them somewhat for the purposes of measurement, usually by means of a measuring device, in order to make a systematic comparison between the object and the device. Measurement is the result of replacing some qualitative aspect of an object or event – its duration, extension, motion or force – by a quantitative estimate, usually pointer-readings on a scale or instrument. In measurement, two quantities are correlated, one set being referred to or compared with the other. Measurement is thus the same as quantitative analysis. Scientific measurement in its simplest form is a more sophisticated kind of classification, as for instance in Dreyer's catalogue of the stars. Tables of chemical elements rely upon measurements but still contain elements of classification. As quantification increases in measurement, the element of classification, which is qualitative, decreases. But the qualitative element can hardly be eliminated altogether.

The standards adopted in scientific measurements are arbitrary in the sense that they are matters of convenience. Measurement of short lengths could be in meters or in feet, for instance; weight could as well be in pounds as in liters. Standards of measurement must be small for small objects, large for large ones. But it would be as useless to try to measure house current in mevs (= million electric volts) as it would be to ask for change for a thousand dollar bill in a small grocery store. The standard of measurement must be chosen from the point of view of its suitability to the size of the objects or events to be measured. Such relative considerations, however, must not be allowed to obscure the fact that once chosen, standards are fixed and all references to them unchangeable.

There are many types of measurement. Roughly, they may be divided into additive and non-additive properties. The additive properties are the most familiar, properties such as weight, temperature and velocity. The non-additive properties are of less interest to science, properties such as surface finish, size or intelligence. We shall confine our interest here to the measurement of the additive properties. Most of the properties of objects can be measured, for instance dimensions, mass, density and shape. Measurements of distance, velocities, accelerations, and of events: complextiy, involvement, reaction, are also common. Measurement calculates the order of objects by numerically reflecting their rank; it calculates also the intervals and ratios between them. Order is sometimes called intensive magnitude; intervals and ratios, extensive magnitude. Rank in the periodic table of the elements is an intensive magnitude, the volume of a gas extensive. In short, measurement takes account of relationships as well as of properties.

Measurements may be simple, as when foot-rules are applied to extended surfaces; or they may be complex, as when Helmholtz in 1852 measured the speed of nerve conduction by electrically stimulating the nerve attached to the muscle of a frog. Measurements may be indirect as well as direct. High temperatures, for instance, of the sort that would melt any measuring rod can be measured by means of the radiation of heat, in accordance with the Stefan-Boltzmann Law whereby radiation increases more rapidly than temperature in a fixed ratio in which the radiation of energy varies as the fourth power of the absolute temperature.

Measurement has logical and empirical requirements which must be briefly noted.

The logical requirement is that every object or event which is to be measured must share with a class of such object or events a common characteristic. It is not possible to measure diverse objects which have nothing in common. Objects not sufficiently alike to be compared for greater than, equal to, or less than, cannot be measured. For example, insurance actuary tables are comparisons of similar life expectancies.

The empirical requirement is that the measure system as a whole must be isolated from the environment, and its parts, which are the objects to be measured and the measuring standard, must be sub-isolates neutral with respect to the measure system and to each other. Of course absolute isolation is impossible of achievement; all actual objects affect and are affected by all other actual things in a falling-off series of importance. The techniques of measurement often involve to some extent the object to be measured. A steel vernier caliper will depress any small object it measures. On the other hand, potentiometers, which are instruments for measuring direct electric currents, do not draw current from the circuit to be measured. The ideal does not exist, but there is such a relation as that of relative non-affectability, a situation in which the object to be measured and the measuring instrument are sufficiently isolated from each other and together form a system sufficiently isolated from the immediate environment so as not to cancel out the accuracy of the measurements taken.

We can now turn to some examples of measurement, first of space, and then of time and mass.

First let us consider the measurement of space. Since all bodies are extended, the establishment of measures of length are very important in science. The history of the attempt to measure length is that of the effort to discover an absolute measuring rod. Of course no absolutes are to be found, and the task is to become as exact as possible. Nothing in existence remains unchanged; there are no rigid bodies nor absolute measuring rods; therefore not only the objects to be measured but our measuring rods as well are in a constant state of flux. There is the additional difficulty of being exact about the task of comparing the one with the other. Measurement is always a business of successive approximations guided by improved methods. The meter length was established as one of the chief measures of length. All standards of measurement are arbitrary, since all lengths exist in nature. The reference for the official meter length was the distance at zero degrees centrigrade between two lines on a platinum-iridium bar at the International Bureau of Weights and Measure, at Sevres, France. Such a standard, no matter how well guarded, was subject to the viccisitudes of war: one bomb could easily have disposed of the bar. Then is was discovered that in the spectrum of cadmium there were two yellow lines exactly one meter apart. There now was a standard which could not be destroyed; it would always be possible to build another spectroscope and again make a spectrum analysis of cadmium. Measuring space is always a matter of comparing length. We measure space by laying measuring rods along bodies occupying space, or by placing them where such bodies could be; we measure, in other words, the space occupied or occupiable as a technique for achieving the necessary degree of isolation of the space we wish to measure.

The measurement of time involves a different sort of measuring rod from that of space. Here we need to adopt some periodic motion as our standard, and any such motion that we adopt is a "clock," regardless of whether it is accomplished by the movement of hands around a face marked off with numerals, the swing of a pendulum, the earth's rotation or some sidereal motion which is uniform. Actually, for the conventional standard of time the solar system was chosen, and the fundamental unit of time is the second, defined as the 1/86,400 part of a mean solar day. Once again, the question of relative isolation arises, and we wish our clocks to be as nearly as possible unaffected by other motions, although irregularities are bound to occur to some extent, there being no absolute measurement of time any more than there is of anything else. The choice of a clock depends very much upon the kind of unit of time we wish to measure: short periods of time such as affect us on the surface of the earth, or long periods of time such as mark astronomical occurrences. The notion of periodic motion as furnishing a clock has been known since Aristotle. It was the difficulty occurring in connection with the translation of local time to other regions which are in motion relative to our own system which gave rise to the problem which was solved by the special theory of relativity. The constancy of the velocity of light in all regions provided a standard for getting from one system of time signals to another.

The measurement of mass involves the notion of weight. Weight is the force of a mass in a gravity field. For purposes of weighing, however, the specific quantity may be considered independent of gravitational pull, since the weighing instrument – scales or balances – is isolated in that it is calibrated locally by comparison with adopted standards of mass. The standard unit of mass in the United States since 1895 has been 453.5924277 grams. Weight is measured by balancing forces, as in the weighing scale or the Wheatstone bridge. The weighing scale is a two-armed lever on which pans are hung at points equidistant from the fulcrum in such a way that the quantity to be measured is related to the known quantity (e.g. numbered weights) by a reading which is zero. The Wheatstone bridge is a branched electric circuit which balances, and so measures, resistances. Other forces are also commensurable on the basis of weighed masses, for instance the tensions on coiled springs and magnetic forces. Tables are available for the estimation of live loads, for instance for various types of materials to be stored in warehouses.

While it is generally conceded that ideas of space and time are derived from the occupancy and duration of matter, mass is not generally understood to be of the same nature and is often confused with matter itself. Mass, however, cannot be isolated and moved about independently of materials. Efforts to define mass have led to difficulties. For while mass is a prime property of matter, matter itself is often understood as that which has mass. Are we when we measure masses, i.e. when we weigh bits of matter, in fact measuring masses or forces which are defined in terms of mass? It may be that mass, too, is an abstraction from matter, that is, from the force of materials. We have said enough at least to see that the measurement of even such elementary properties involve a further step in abstraction, and when we learn that advanced stages of science deal with the measurement of more complex processes, we are safe in concluding that science is concerned primarily with certain sets of abstractions, which are no less abstract for having been derived from concrete matter.

Progress in science involves improvement in the techniques of measurement. Trigonometry arose as a measuring technique, although, like all mathematics, it is independent of the practical uses to which it may be put. We might generalize this example, and say that the largest part of applied mathematics has in some way or other to do with the techniques of measurement. Differential equations, for instance, are methods of measuring accelerated motions. Statistical probability measures groups of data. Combinatorial analysis is now being tried to measure economic relations of exchange.

The use of mathematics itself involves a further stage of abstraction beyond that brought about by the hypothesis and the instrument. For example, by means of interpolation the investigator replaces the values which he has measured, and which are discrete and discontinuous, by a continuous function which as such has no direct representation in nature. By means of tables of values he arrives at the notion of mathematical function which is similarly abstract. Thus the data measured suggest formulas which in themselves are abstracted to degrees no longer exhausted by the data.

7. *The use of techniques*

A scientific technique may be understood as a special way of operating in science. It requires the acquisition of skills as well as the use of instruments and the manipulation of materials.

The scientific method is above all an operation which cannot be performed by just anyone; it requires special skills, and the skills presuppose practice and adroitness, expertness in action: precise response built into the proprioceptive system by habitual behavior. The stages of the scientific method constitute broad outlines, which need to be filled in with greater specificity and particular adaptation to the problem at hand. An experiment may require a treatment comparable in nicety only to a piano performance of Beethoven, if it is to be successful. Thus manual dexterity often must be included among the prerequisites of the scientific techniques.

Every science has a different field of operations, and though these are all related they do differ in many important respects; in complexity, accessibility, and degree of penetrability, for instance. Scientific techniques must therefore of necessity vary from science to science. Among the techniques to be found in chemistry are: solution, filtration, evaporation, distillation, and crystallization. None of these would be appropriate in astronomy or psychology, for example. On the other hand, the double-blind technique, whereby the drugs and placebos which are placed in the hands of a practicing physician for trial runs are not disclosed to him nor to the patients, thus insuring an additional protection against a weighing which would adversely affect the success of the experiment, is commonplace enough in biology but would not work at all in physics.

(a) *Concrete Models*. One technique in general use consists in the building of models. Models may be concrete or abstract, that is to say they may be material or mathematical. Let us say a word about each of them in that order.

First the concrete models. Concrete models are material analogies. The mechanism in nature is assumed to exist, from the phenomena observed. Then it is compared with a working model constructed artifactually. The model may be smaller than the original, as for instance the clay model of a dinosaur, or it may be larger, as for instance the wooden model of a chemical element. Any structural representation is a model, and may consist merely in a drawing on paper, such as an architect's blueprints; or it may be a solid, such as the "pilot model" used in planned mass production in industry.

The use of models can influence research. For instance, if a model be adopted however tentatively, further investigation may take the direction of exploring the unknown properties suggested by the model. Models in science serve two experimental purposes. They may be used for illustration or for demonstration.

First as to the models used in illustration. These are concrete examples of abstract ideas. At the turn of the century, investigators who were to take part in an expedition to observe the transit of Venus were trained to observe a mechanical model placed at some distance from a telescope. The orrery, a model of the bodies and movements of the solar system, is even older. Constructions illustrating configurations in geometry have long been familiar. The best known are the polyhedral models so often found in science museums and university mathematics departments. Models in biochemistry illustrating the relationships of atoms within the molecule and in biochemistry of complex compounds are now common. The scale model of the complex protein molecule, deoxyriboneucleic acid (DNA) built by Watson and Crick, is a good example.

(b) *Abstract Models*. There is a sense in which the word, model, is used to describe a large mathematical theory concerning the structure of some natural phenomenon. One way to test an hypothesis is to construct an abstract model corresponding to the hypothesis in order to see by means of "a thought experiment" whether all of the relevant data could be fitted into it. There was a time when Newtonian mechanics was the only model for macroscopic physics besides Newton's theory of gravitation and his calculus. But then the model had to be modified because it could not account for fast particles as well as relativity mechanics or for small ones as well as quantum mechanics. The scale models of the atom did much to facilitate the progress of research.

Every hypothesis is capable of being given a certain measure of verification by an unlimited number of interpretations provided that they are not those from which it was originally derived. This is the use of models in demonstration. Models which provide evidence for hypotheses may be physical models, but they may also be imaginative constructions. A dynamo is a mechanical model of certain principles in electromagnetics. Poisson's model of the ballot-box is an imaginative model of the probabilities of an event. Bohr's model of the atom served to support a particular interpretation of quantum mechanics.

The risk in using models is that the model will be mistaken for a graphic representation of the theory it follows. For a model is only a partial representation even when it is a true one; rarely does a model contain all of the features of the object but only some of them, usually only those that are known where others as yet unknown believed to exist. Models are at best only approximations, and it is enough if they do suggestively the little that they do. For even in this case there is some unavoidable degree of distortion involved in the mere selection of the known features. In a true model of all the features, those which are now known have to take a different position of relative proportion.

8. *Experimenting*

In the conduct of an experiment, there are four well defined steps.

The first of these may be called the preparation of the specimen. Accessible material is usually (thought not always) brought into the laboratory where it can be properly isolated and got ready: thin sections cut by means of microtomes, slides prepared, control groups selected. Field experimenting requires the choosing of the proper time and place. In geology many hours of exploration are necessary in order to discover exposed strata of the type desired. In the case of inaccessible material, such as astronomical bodies, the time and place for experimental observations will have to be carefully calculated, such as is done before expected eclipses. The British expedition to South Africa led by Eddington immediately after World War I to verify Einstein's prediction that light rays would bend close to the sun's surface, is an instance in point.

We have prepared the material to be tested, but what about the apparatus which is to be put to work on the material? The second step involves the preparation of the instrument. There are tasks peculiar to it quite apart from what it may be testing or what may be the results from running a series of experiments. There must be a checking of the equipment before the start of the run. Instruments not in continual operation have to be prepared to operate. Parts must be cleaned, motors tuned. Push buttons, toggle switches, knobs and levers have to be in good working order. If there are tubes, they have to be given time to warm up; if there is a vacuum, several days may be required to pump it up to the required level. Is everything properly oiled and tuned up? Do all meters, indicators and gauges read as they should? Is the aparatus operating properly? Perhaps a dry run is in order here; the instrument can be turned on without the material to be tested just to be sure that all of its parts are in working order. For this purpose a check-list is a good thing to have. Check points can be itemized and used as a reminder for a thorough examination of the apparatus before an experiment is run.

The third step consists in the proper isolation of the experiment. The material to be tested and the apparatus used to do the testing must be removed as a unit from all factors which might constitute interference. Factors ignored or held negligible have the power of cancelling the effectiveness of an experiment. Variables which are not controlled as parts of the experiment must be properly randomized in order to neutralize their interference, for instance by experimenting at different times and if possible also at different places. Randomizing unwanted variables is not an ideal way of eliminating the danger which they constitute, but it is one way of reducing it.

In the case of experiments in the field rather than the laboratory, the third step consists in preparing the site and arranging the conditions under which observations are to take place. When Tinbergen wished to observe the social life of the herring gull under average conditions, he had to prepare his observation site well in advance of the events he had chosen to witness, and he had to spend many hours in concealment in order that the birds might learn to proceed normally in his presence. The behavior of the objects under observation, whether it be the nesting habits of the Kittiwake on the high cliffs of the Farne Islands off the coast of Northumberland, or the occulting periodic of Cepheid veriable stars, requires many hours of vigil and prolonged observations.

The fourth step consists in the running of the experiment. Once the experiment has been begun, the operation of the instrument needs to be watched. Maintenance during the run is important, and in this connection a maintenance check-list may be made up and followed, with oil check-points, dials, levers, etc., listed. There must be no failure of performance which might color the findings. Disturbances arising from inside the apparatus due to fluctuations in its source of power or to increase in noise must be anticipated wherever possible and controlled. Tubes can always blow, circuits go out, ventilation can become clogged, liquid flows can become turbulent, transformers can produce undesirable magnetic fields, moisture can form on cooling coils, mechanical systems can develop oscillations, motors can show vibrations. Such a list would have to include provisions for the aging and wearing of parts as well as for operationally developed maladjustments. These and many similar difficulties can often be anticipated and avoided, though of course not always. If not, they must be met when they occur, and corrections made as the experiment proceeds.

An investigator who is familiar both with the hypothesis he purports to be testing and the construction and operation of the instrument which was designed for the purpose has a better chance of successfully carrying out the experiment than one who is ignorant on some one of these scores. A great deal of fortitude in the face of obstacles and persistence in the face of failures is necessary for the success of some experiments. So much can go wrong, so many little things may have been overlooked in a first run, that the project may be abandoned before it has had a fair chance. Instruments have been discarded for a time, only to be returned to favor at a later date after the occasion for their rejection has been dissipated. Such was the case with Carrel's perfusion pump, which was given up after Carrel's defection to the Vichy French during World War II, only to be taken up again by the biochemists a decade later.

A certain level of sophisticated observation is essential during the running of an experiment. In every experiment there are many conditions to be noticed, as it were, all at once. The various parts of an experiment must be watched and their coordination maintained in order to preserve the continuance of their proper interrelations. If there are relevant factors which may be varied, they must be varied one at a time. Irrelevant factors may be varied to be sure that they are in fact irrelevant. Positive efforts must be continually made to ward against the inadvertent introduction of elements which may constitute serious interference.

An astute experimentalist will be alert to the failure of an experiment. A failure must not be confused with a negative result. An experiment which fails is one which is incapable of giving any result, as for instance when an apparatus breaks down or when there has been found to be sufficient interference to cancel the success of the experiment. Changes in the behavior of materials which are parts of the apparatus when under a high vacuum, immense pressure, enormous speeds, or extremes of heat or cold, will have a different meaning from those same changes in the materials which are the subject of experiment. On the other hand, consistency of policy must be maintained: once a factor is judged to be an interference, it must be so judged on all subsequent runs of the same experiment.

The fifth and final step in experimenting calls for the collecting of results. In the case of laboratory experiments automatic recorders, rate meters, counters, scalers and timers must be read carefully, and in the case of field experiments, observations must be recorded. Both should be done promptly, before a lapse of time allows inaccuracies to creep in. The experiments were performed for the sake of the results, and the results will have to be interpreted; but collecting results is often a procedure requiring considerable promptness and great care.

Generally speaking, observation at the experimental level is no longer naive, and the investigator comes to his observations with deliberate preconceptions. He knows what it is that he should expect to see: a movement, a configuration, a pointer-reading, indications as significant in their absence as in their presence. Negative results must be carefully noted; as we have seen, a negative result is not the same as a failure: a properly designed experiment can be successful in yielding either a positive or a negative result. It must not be in science, as Francis Bacon said, that men mark only when they hit and not when they miss. An experiment is the narrowing of the evidence to the relevance of a certain question; however, there is always the possibility of surprise to be kept open.

9. *Types of experiments*

A typical experiment will consists in two events causally connected. Although the two are properly parts of one and the same experiment, at the same time if they are not properly distinguished they cannot be properly related. Something is done and something else is the result. Given the variety of materials to be tested and the number of kinds of investigative apparatus it is possible to construct, there is no hope of specifying all of the types of experiments. Their richness and variety is evidently inexhaustible. All we can hope to do is to suggest something of what may be encountered along the way, the difficulties and also the advantages.

An experiment is an attempt to penetrate beyond ordinary types of experience. Observations with instruments do this and so do experiments. In any advanced science, experiment does not take place at the level of ordinary experience but at high analytical levels. We do not experiment with sticks and stones but with cells and molecules; we do not measure ordinary speeds and distances but the speed of microseconds or of light-years. The behavior of an individual is not of as much interest as the behavior of large groups of individuals, larger than we could observe with the unaided senses. Mathematics performs operations for us that we could not perform for ourselves, and it takes us into regions of abstractions where we could not otherwise have penetrated and from which we return to the concrete world with enormously effective techniques of applicability not otherwise available.

Every new advance in instrumentation makes possible a corresponding advance in the type of experiment. The present group of instruments engaged in nuclear research, the big accelerating machines designed to hurl particles at each other at a speed calculated to disintegrate nuclei, the synchrotons, and the linear accelerators, have achieved new depths of penetration. And the same things can be said for special biological methods; for instance the radioactive tracers and the new tissue culture methods. New methods of measurement are often the result of using old instruments in new ways. Scintillation counters can now be adapted to determine the degree of natural radiation of living organisms. Exceedingly fast electrons are scattered by atomic nuclei in ways which make possible the measurement of the nuclei. X-rays are used to measure the strength of metals.

There are no fixed rules for making fresh advances in experiments or for breaking through to novel lines of endeavor. Galileo opened up enormous prospects of original research when he made a new instrument: the telescope. Jansky invented radio astronomy in 1932 when he used an old instrument (the antenna) in an entirely new way (to detect stellar radiation), thus adding an auditory telescope to Galileo's visual one. Berger put an old instrument to use in a different science when in 1928 he attached electrodes to the human brain and found that it gave off electrical impulses, thus initiating the study of electroencephalography.

Some hypotheses become increasingly confirmed as diverse types of experimental approach are directed toward them. Two distinct methods of measuring the binding energies of nuclei have yielded the same results: one method a direct measurement and the other an indirect one by means of the measurement of mass defects, thus offering firm support for the hypothesis concerned with the measurement. Avogadro's number for the molecules in a gram-molecule of gas under ordinary pressures and temperatures (somewhere between 6 and 7×10^{23}) has been confirmed in three different ways: by investigating the scattering effects of light by gas, again by studying the emission of X-rays and finally by the distribution of radiant energy in thermodynamic equilibrium. The double or – as in the case of Avogadro's number – the triple agreement of experiment with calculations certainly does offer convincing confirmation.

Experiments in which control groups are used are quite common in biology. A controlled experiment is an experiment in which all factors are kept constant except the one under investigation. In the use of a control group, the specimens to be examined are divided into two groups which are equal in all respects, and one is disturbed by the introduction of an extraneous element. The control group is the one left undisturbed, while an experiment is conducted on the other. The effects of the disturbance are observed by a comparison of the two groups. The choice of the condition to be varied while the others are kept constant is not inherent in the nature of the specimens but is made by the investigator in terms of some hypothesis he wishes to examine. And in making the decision to conduct the experiment in this way precisely, he has acted from a deduction from the hypothesis or from an anticipation that the results of the experiment will inductively arrive at the hypothesis.

Let us suppose for example that there has been a reason for determining the effect of a variation in temperature on the health of rats. For this purpose two groups of rats are selected, which must be as alike as possible: rats from the same strain and even from the same litter of from different litters with the same parents must be represented in each of the groups. The groups are placed in identical cages, glass enclosed and airtight, in which the temperature can be controlled. Now all conditions in both cages are maintained in exactly the same way, including kind of food, time of feeding, light intensity, and so on. The independent variable, which is the experimental factor, is: temperature. In cage *A* the temperature is maintained at a constant 70°F., while in cage *B* the temperature is systematically varied from 40°F. to 80°F. After a suitable interval of time, the results are observed. The rats in cage *B* are found to be livelier, they eat more, their coats are glossier, they weigh more, etc. It is therefore concluded that the hypothesis that variation in temperature is a desirable stimulant to the well-being of rats.

A great variety of types of experiments exists and others are being added constantly. The artificial reproduction of natural entities tells the investigator much about those entities. A good example is the synthesis of protein recently accomplished, and the same is true of the slightly older transmutation of atomic particles. Experiments are undertaken for all sorts of reasons, for there are many purposes served by experiments within the range of the scientific method. Experiment are undertaken for instance to eliminate errors of observation, to ascertain the character of an area to be investigated, to probe a subject-matter whose systematic investigation is contemplated. The bombardment of nitrogen atoms by alpha particles often produces the emission of hydrogen nuclei from the nitrogen nucleus. Rutherford in this way discovered the proton. The experiment was crucial to the theory of atomic nuclei, and an entity discovered that was later investigated in other ways.

Unexpected results uncovered in the course of experiments compel a change in the hypothesis as often perhaps as they verify or falsify it. When an investigator decides to run an experiment, he exposes himself to the discovery of results which cannot be altogether predicted. He cannot control all the conditions; if he could, he would be calculating rather than experimenting. To experiment means to this degree to invite the unknown. Rayleigh discovered argon because there was a discrepancy between two samples of nitrogen, the one derived from air and the other produced by chemical methods. The chance discovery of radium by the Curies and of penicillin by Fleming in the course of work aimed at quite different directions are well-known instance of chance discovery.

Lastly, perhaps, should be mentioned the effects of imaginary experiments which have been counted on in calculations without ever having been performed. A better name for this type of experiment is "thought experiment." When it involves the use of the observation of graphs, as it often does in mathematics, Peirce called it "ideal experimentation." Dalton's atomic theory of the chemical elements announced for the first time in 1803, was built on the work of his predecessors, notably Priestley, Cavendish, and Lavoisier, who had of course conducted laboratory experiments. Dalton in this particular connection did not, but he did make his own spherical pictures of the atoms and he rested his theory on them. The contraction of bodies at right angles to their direction when travelling at speeds approaching the speed of light in a vacuum, contained in the speculations of Fitzgerald (the "Fitzgerald contraction"), was employed by Lorentz to explain the failure of the Michelson-Morley experiment to detect the ether drift. In certain cases of advanced work, the confirmation of experiments which for practical reasons cannot be carried out may be assumed in order to use the results in connection with experiments which can. But this procedure must be used judiciously, and there must always be a performable experiment in the complex.

The theory of relativity was suggested to Einstein by the experimental findings of others, and although he himself was no stranger to the physics laboratory he did not carry out any laboratory experiments in connection with this particular theory. It only goes to show what "thought experiments" conducted under the proper circumstances by laboratory-oriented scientists can accomplish.

10. *Varieties of results*

The completion of experiments to test hypotheses produces a certain result. If negative, that ends the inquiry, but if positive it consists in the discovery either of a significant datum or of groups of significant data. The attainment of such results is of course the entire purpose of the experiments. It is understood of course that if the experiments are successful and the yield positive, the matter does not end there. The hypothesis goes on to other types of testing before becoming firmly established as a law. But there are interim conclusions to be considered, and we must look at these.

First, then, as to the significant datum. Hypotheses are usually universal propositions having an empirical reference, and they can accordingly be supported by experiments. The experiments themselves stand as propositions in action, and they count of course as single instances. The material object is a datum which has been discovered by means of experiment and it can be a sub-atomic particle for instance, such as a brief-lived meson, or it can be a large-scale macroscopic object, such as a new planet. It can be a gas, a liquid, plasma, or a solid. There are in addition hypothetical entities which experiments show must be set up which are not believed to be actual: limiting cases, such as perfectly frictionless engines or ideal radiating black bodies. Such hypothetical entities are called "empirical ideals," as we have already noted in Chapter III (3), but they differ from hypothetical entities which refer to supposed material objects in being ideals and therefore of necessity non-actual.

In general the data can be classified as general types of material objects with their properties, and of the relations between them. Given the interconvertibility of matter and energy, which seems well-established, we shall have to include types of processes as well as of entities among material objects and also some important intermediate types, fields and forces, for instance. The relations between types of material objects can be various: all the way from statistical tendencies to causes, from correlations to continuous functions. It will be well to say a word about each of these. But first a word about properties.

There is small difference indeed between material objects of one of these sorts and the properties of material objects. Chlorine is a greenish yellow gas; it can provide the atmosphere for burning active metals; it combines violently with hydrogen in the presence of sunlight; properties which were revealed for the first time by means of experiments. Again, some of the unicellular organisms, called algae, can swim by means of *flagella* (their proto-plasmic tails). Spectrum analysis or methods of chemical assay are attempts to discover certain types of properties of elements.

Processes are events in which there is a rearrangement of the parts of a material object or in which there is an interaction between material objects. Studies of beta disintegration or of cascade decay in subatomic physics, of the mechanism of heart action or of kidney function in biology, are examples of processes revealed by means of experiment.

Lately, intermediate types of phenomena, classifiable clearly neither as entities nor as processes but partaking of some of the characteristics of both, have come into prominence. These are forces and fields. Forces such as electromagnetic forces, gravitational forces and nuclear forces are dealt with in physical experiments. There are fields corresponding to each of these forces. A field is a neighborhood in which a force of a particular kind is exerted. In general, a field exists at one analytical level lower than the entities which are characteristically affected by it.

Broader than interactions between material objects are the functions of various sorts, from mere correlations to the simple two-valued functions in which there is one independent and one dependent, to the more complex functions such as the logarithmic functions or continuous functions, or the functions of a complex variable. Functions are expressed as equations, such as the partial differential equations of modern physics; but the material counterparts which call out the mathematical language and bring about the need for it are indicated by experiments.

Broader still than functions are those tendencies of groups of entities or of processes roughly described as statistical probabilities. Extended and repeated experiments are often necessary to study the statistical features of groups which are not found in individuals, such as are uncovered in learning experiments with laboratory animals, such as mice or hamsters. Strong impressions of this sort point to causal relationships. This is still a disputed area in scientific investigation. It has not been possible to refute Hume's point that causality cannot be made empirically evident. But then much of what passes for experimental evidence is based to some extent on inference from what is empirically evident. Whether there is or is not such a thing as a parameter behind every statistical probability, as Fisher, for instance, would have us believe, the fact is that experimentalists proceed as though causality does exist.

No doubt new types of entities and processes, new types of material objects altogether, will continue to be discovered by experimental means. There was a time in the past when fields and field forces were unheard of, and in the future there will be structures at present not envisaged come into our knowledge. We must in science always leave room for fresh developments. What kind of things will future scientific research encounter? We recognize now various kinds of laws and diverse relationships. Others are sure to follow as current conceptions of discreta and continua are intermittently challenged and modified. The varieties of data have no more been altogether explored than has anything else in science.

11. *Interpretations of the data*

Interpretations of data resulting from the successful completion of experiments may be of two kinds. They may consist of interpretations of a single significant datum or of a few data, or they may consist of interpretations of large collections of data. The methods of interpretation employed in the two cases are very different. We shall treat of the first in this section and of the second in the next.

Every stage of science has its own inherent difficulties and requires its own peculiar kind of intuitive understanding, and there is no exception. From the original observations through the discovery of hypotheses, the design of experiments, the conduct of experiments and the interpretation of the data, each requires its own type of brilliance and has its own peculiar subtleties; each, too, has its own style of error in terms of which all of its possible effects can be cancelled. The fact that an experiment has been conducted in order to test an hypothesis and has concluded in certain results does not necessarily mean that the relation of the data to the hypothesis is easy to determine. For, as we have already noted, it is seldom that the results are conclusive for the hypothesis; they either support it to some extent or give evidence against it in some way. But to what extent and in what way, it is often extremely difficult to decide. The investigator has to bear in mind the tentative and approximative nature of all empirical verification. In propositions of a general nature which are supposed to be supported empirically, as is the case with most theories in science, the investigator does not think in terms of proof but rather of degree of verification. Then again, he has to consider for every hypothesis the range of possible explanation of the data which have been gathered for it by means of planned experiments. How many possible interpretations of the results could there be? Do alternative explanations present themselves to him? If so, how is he to choose; or, in other words, how is he to determine adequacy? A number of conclusions may fit the data, though not all equally well.

The problem, then, is to decide which conclusion fit best. A common error is to suppose that because a certain conclusion seems to fit better than others that it must be the obvious – and only – conclusion. Meanwhile the existence of another conclusion which may fit still better remains unknown. It is quite possible for scientists to overlook the fact that observation is of data only: all the rest is interpretation. So much of what is found as data can be interpreted in more than one way. Thus there must be theoretical considerations lying outside the scope of the experiment itself which determine to what extent the investigator is to take the experiment as verifying or falsifying the hypothesis. Thus it often happens that every step in the scientific method is carried out correctly except the last, but this can be sufficient to nullify all the previous efforts. Although there was common agreement up to this point, the question of the correct interpretation of the data can still divide investigators.

This is not to say that arguments will center upon the data; they can be agreed upon and still give rise to controversy; for it is not the data themselves but what the data mean that offers so many difficulties. It has often been pointed out that the data themselves can never be false; they are whatever they are. Whatever has happened has happened, and there can never be any way of altering that. But the investigator can make the wrong interpretation of the data. Even though he approach the data as free as possible from preconceptions, he can still make mistakes concerning the relevance of the conclusions to the hypothesis they were intended to test. He may misread the data or wrongly evaluate them. Remember that no matter how much evidence is accumulated experimentally in favor of an hypothesis, it constitutes no proof. Proof in these terms would involve the complete exhaustion of all possible instances, and this, in view of the range of present instances and the inaccessibility of past and future instances, is entirely impossible. Then how many instances will be required to satisfy investigators as to the truth of an hypothesis? There are no quantitative standards established which could serve as references for the discrimination of verified from falsified hypotheses. Thus we speak of *evidence for* or of *evidence against* but never of absolute proof or disproof.

No number of instances of verification will render an hypothesis certain. Experimental verification is a matter of degree, and the probabilities of the hypothesis are directly proportional to the strength of the experiments. We shall discuss this last problem in the next two sections; meanwhile it is important to remember that certainty belongs exclusively to the domain of logic. There exists no well defined decision procedures for determining whether according to the experimental evidence a given hypothesis is verified or falsified and in either case to what degree. At the present time, far too many factors are involved, and whether they are controlled variables, randomized variables or uncontrolled variables makes no matter; for the decision is still a matter of expertise and not of any formal rules. It is to be hoped that eventually rules can be found and established to guide the experimental scientist. As we shall see in the following two sections, the investigator is faced with multiple choices which complicate the nature of his difficulty.

What constitutes sufficient evidence for or against an hypothesis? Should the modification in spectral lines emitted by an atom when passing through a uniform magnetic field (the "Zeeman effect"), which had been used by Lorentz but which could be explained neither on the basis of his electron theory, the original quantum theory or wave mechanics, have been considered sufficient support for the hypothesis of the existence of the electron, as in fact it was? The answer is in the affirmative, though it should be noted that such single strands of evidence are not often accepted in other sciences. The degree of assent to an hypothesis must be in proportion to the weight of evidence, and weight of evidence is not determined by a simple counting procedure. It may often happen that a single crucial experiment, such as the one carried out by the Dutch physicist, Zeeman, in 1896, carries more weight as evidence than some tremendous but insignificant collection of psychological statistics. Thus the investigator must learn to have weak or strong hold on beliefs, as though there were degrees of probability of the truth of proportions, rather than entertain firm acceptances or rejections.

When an experiment intended to test an hypothesis verifies it, there are still grave matters of interpretation awaiting. A few of the possibilities which arise in this connection may be mentioned. It may happen that the evidence is so strong as to constitute definite support for the hypothesis: that the evidence is so weak as to be barely sufficient to deserve the name; that the evidence may be strong but partly (or totally) irrelevant; that there were unknown factors present in the experiment which weaken its value as evidence; that the evidence while verifying the hypothesis reveals serious weaknesses in it, as occurs for example when the hypothesis is shown to have a smaller than supposed range; that the evidence may seem strong yet actually be weak, because of the poor design of the experiment. These are some but by no means all of the considerations which may have to be weighed in estimating the results of a successful experiment.

There is also of course (and equally important) the case when an experiment intended to test an hypothesis falsifies it. Such a result also may mean a successful experiment, and similar matters of interpretation are raised. The experiment may falsify the hypothesis, but also there may have been errors in the conducting of the experiment or in the framing of the hypothesis; or they may have appeared what Herschel called residual phenomena: unexpected discrepancies due to unknown and perhaps irreducible minimal elements in the experiments. The same difficulties that were recited above in the case of the verifying experiment can be recited again *mutatis mutandis* for the falsifying experiment. There are in addition instances in which each of the types of results are capable of interpretation in terms of the other; it may be stated here for the falsifying experiment: sometimes the refutation of an hypothesis is not sufficient to falsify the hypothesis; instead, it helps to show the conditions under which the hypothesis holds true.

The experimental decision is crucial with respect to the hypothesis. After an experiment or a series of experiments verifies an hypothesis, it is passed along to the next procedure, and there are very definite steps to be taken. But when an experiment falsifies an hypothesis that hypothesis must be abandoned, and the investigator must prove himself as willing to set it aside as he was to adopt it for purposes of inquiry in the first place.

Immense logical skill and a remarkable sense of proportion is required to determine the type and extent of the relation between the hypothesis being tested and the data which result from the tests. The commonest error in this regard is to suppose that because a conclusion has been reached from certain data it must necessarily follow from those data. It often happens that an investigator will suppose that because his conclusions were arrived at as the result of prolonged, arduous, and often ingenious observations and experiments, the conclusions must for that reason be incontrovertible, so that to question them would be the same as questioning the data themselves. This is only too often far from the case. For before it is fair to regard the conclusions as being as sound as the data, they would have to be shown to be the only conclusions that could be come by as the result of examining data. And how often is this so? Most conclusions are the results of inductions from data, and since when are inductions logically necessary? The only inductions which can be said to be necessary are those made from total collections of data, not ever those made from samples; but such inductions are indistinguishable from deductions. Inductions made from sample collections of data are never more than statistically probable.

The interpretation of data sometimes calls on the astuteness of the investigator to recognize in them the verification of an hypothesis or further support for a theory already verified. It could not have been easy to have recognized, in the fact that electrons accelerated to enormous speeds by the magnets of the cyclotron gain in mass proportionately, a striking confirmation of Einstein's theory of relativity.

Interpretations of data of course change as new experimental evidence is made available. Lavoisier as the result of his researches defined a chemical element as a substance which cannot be decomposed by any known means; which was true for his time. But as more means became known, the molecule, the atom and the constituents of the atom were all recognized as products of the decomposition of chemical elements. As the scientific method obtains new instruments and techniques, old theories become falsified and new theories receive the firmness of facts. The Ptolemaic theory of the geocentric universe is a thing of the past, and the theory of electrolytes: that acids, bases or salts may be decomposed by the current they conduct by means of ions, is no longer a theory but a fact.

It often happens in experiments that we can know the effect of certain operations without at all understanding the processes. This is true, for instance, of the so-called photo-electric effect of radiation on matter where we know what happens but not how it happens. Again, Miss Leavitt discovered the period-luminosity relationship in stars merely by examining in 1905 a collection of photographs of the Magellanic Clouds and discovering the peculiar occulting properties of one class of stars which were later named Cepheid variables.

The history of a significant experiment is that first its significance is noted; then the experiment is repeated many times, with additional refinements of technique; and finally the experiment fails to operate in more complicated situations, thus forcing a reconsideration of the problem which the experiment was presumed to have settled. The experiment of Dulong and Petit in 1819 to devise a method for determining the specific heats of elements was carried out on mercury. In 1851 a refined version of the experiment was carried out by Regnault with more exact results. The relation of this experiment to Avogadro's Law was noted by Cannizzaro in 1858. Subsequently, the method followed in the kinetic interpretation of specific heat was similar to that of Dulong and Petit; but important exceptions began to appear when complex molecules were introduced, for interatomic forces do not allow the molecular heat of compounds to follow the Dulong and Petit relationship.

The discovery of an impossibility has often disproved an hypothesis only to suggest another and more important one, as in the case of ether and the special theory of relativity. To disprove an hypothesis, it is often necessary only to derive a single consequence from it and then to show that this consequence is contrary to fact, as in Kant's example of the polestar which would always appear at the same altitude if the earth were flat but which is in fact not always at the same altitude. However, not every collision between fact and hypothesis is fatal to the hypothesis. It may be that a simple revision of the hypothesis will be sufficient to re-establish the required consistency; or, when other facts of the same class are added to the fact already recognized, a probability in favor of the hypothesis can overweigh the single fact against it. The decision as to whether a fact does or does not count fatally against a relevant hypothesis is a nice question and often cannot be settled except by the kind of judgment which relies upon the similarity to other such decisions previously settled successfully.

The interpretation of the data is made no easier by the fact that the laws of science are never absolutely fulfilled. There is always a surd element, an extenuating set of circumstances, which must be taken into account in estimating the results of any experiment. Thus the rejection of an hypothesis in terms of a single exception is often dangerous and misleading. But "exception" must be carefully distinguished from "extenuating circumstance." Exceptions are found in statistical laws; for instance, if carcinoma of the lungs kills 87% of those who contract it, then the other 13% constitute an exception. Extenuating circumstances are found in all causal laws. Gravitation so far as we have discovered holds without exception; yet since it is stated ideally for a vacuum, and there is no perfect vacuum, the atmosphere through which bodies attract each other is always specified as an extenuating circumstance.

Another difficulty is that the data are not always uniquely determined. There is the famous case in subatomic physics of the wave-corpuscle duality, whereby both light (photons) and matter (electrons) under some experiments exhibit the properties of waves and under others the properties of corpuscles. In the case of the gas laws, we have the duality of continuity and discreteness. The controversy over the wave-corpuscle duality, which some investigators, like Heisenberg and Bohr, would like to interpret in the form of a principle extended to the whole of physics, as in Bohr's principle of complementarity, but which some investigators, notably Einstein and DeBroglie, Vigier and Bohm, would like to assume marks the limitation either of an instrument or of an investigation in a given direction by certain means, is an excellent example of the difficulties confronting interpretation in many cases where the data are agreed on.

Support for hypothesis sometimes comes not from the accumulation of evidence but rather from the agreement of disparate evidence. There were three possible interpretations of the negative results of the Michelson-Morley experiments of 1887 and after, designed to detect the drift through the ether: (i) there is no motion of the earth through the ether; (ii) the ether moves with the earth; (iii) there is no ether. Einstein's decision to rely on the last of the three was dictated by the fact that he was dealing at the same time with other data: with the Fitzgerald contraction, with the Lorentz transformations, and with certain new mathematical equipment: with Maxwell's equations and with Riemann's non-euclidean geometry. And somehow these harmonized more easily with the third of these interpretations than with the other two.

Experiments are designed to test hypotheses. They may in addition to this give fresh or more precise results. Hypotheses may be not only verified or falsified by experiment; they may also be improved. In certain cases, hypotheses are neither verified nor falsified by experiment, or if they are, this is not the most important result of the experiment. They may be true only under certain conditions; they may be false and suggest the direction in which truth is to be found. Hypotheses often have to be revised in and by their verification. Prout's hypothesis that all atoms are formed from hydrogen had to be revised to the theory that all atoms are formed from the constituents of hydrogen.

The nature of evidential support for one kind of hypothesis may have to consist in showing that some events in the world would never have happened were the hypothesis not true. These indirect methods are often valuable. The course of scientific development is a tortuous one. A tendency that shows up in the experiments may be significant or it may mean merely that some variable has not been brought under proper control; that samples may not be sufficiently random; or that hypotheses (more often than not) simply prove to be wrong. Not all hypotheses that are abandoned, however, were invented in vain; many make their contribution to scientific advance through the power of suggestion. This is true to various degrees of the phlogiston theory, of the theory of the ether, and of Rayleigh's law of spectral distribution.

In this section we have been concerned throughout with what are perhaps the simplest cases of the interpretation of data. Needless to add, more complicated cases exist. And in addition, there are all sorts of variations. One may be mentioned in illustration. It often happens that experiments are carried out not to test an hypothesis but to decide between rival hypotheses. Estimates of the brightness and distribution of remote clusters of galaxies may serve to decide between the theory of cosmogony proposed by Lemaitre involving the evolution from a primeval atom and the newer steady-state theory of Bondi, Gold and Hoyle.

The end products arrived at by interpreting the data of experiments are verification (or falsification) for the hypotheses involved. An hypothesis which was the inductive result of certain observations suggested the design of experiments which could support or disallow it. The experiments were carried out accordingly and yielded certain data. The data were interpreted in the light of their relevance to the hypothesis they were designed to test. Now the investigator is led either to the abandonment of the hypothesis – more tentatively, perhaps because of his establishment of evidence in that direction – or to the strengthening of the hypothesis so that further investigation into its validity seems justified. In the latter case, he begins to speak in terms of a theory, for a theory may be defined as an experimentally confirmed hypothesis. The explanation of what that means must wait until the next chapter.

12. *Empirical probability*

The statistical method is one aspect of the scientific method, but the scientific method is not exclusively nor necessarily the statistical method. There have been fully developed instances of the scientific method in which statistical calculations did not occur, Newtonian mechanics and Darwinian evolution, for instance. A sharp distinction must be drawn between testing hypotheses statistically and testing statistical hypotheses. Some experiments are statistical, some are not, independently of the type of law tested. When an investigator talks about scientific inference, he often means some method of treating masses of experimental data whereby an hypothesis is to be tested; he does not mean the discovery of hypothesis, for this is rarely statistical. The data which can be processed by computing machinery belongs to the experimental stage of scientific method and not necessarily either to the first observational stage from which inductions are framed or to that later stage of calculation from which data are predicted by means of probabilities. Before data can be collected, a decision must have been made concerning the canon of collection, and for such a decision an hypothesis of some sort was necessary.

It should be noted at the outset of the consideration of statistical probability that the process is at no stage mechanical. Computers may be employed, and often are, to collect, sort and store vast accumulations of data, but they never make the decisions involved. Inductive procedures in so far as they initiate projects or affect decision-making are always the work of the investigator himself. Statistical inference is involved in the choosing of alternative hypotheses on the basis of the evidence but never initiates that choice.

Statistical probability belongs to the experimental stage of science as one variety of result in need of interpretation. Prior probability is theoretical, posterior probability experimental. The most formidable and far-reaching theories in science have not been statistical in character nor the results of experiments involving huge masses of data. An induction to an hypothesis is the same whether derived from one instance or many, because probability does not change the character of the hypothesis. Every law of science rests on inductive evidence, in the sense that it is prepared to withstand the experimental tests. For example, in every instance that has been observed, a body on which no external force is acting continues either in its state of rest or moves with uniform motion in a straight line (Newton's first law of motion). Presumably, bodies so isolated will continue to behave the same way in the future. The accumulated evidence is overwhelming.

Most statistical inferences function for or against hypotheses, and the hypotheses themselves could be causal or statistical – absolute causal laws or statistical inferences. It is perhaps the rare crucial experiment yielding a single result which proves to be of the most significance. This is very different of course from predicting a specific event. There is no relevance to the prediction of a single instance; statistics can predict only statistics. Most frequently, however, experimenters are confronted with the weaker situation presented by large accumulations of data. Probability occurs experimentally in the testing of hypotheses when those hypotheses are stated formally. The probability of the truth of Einstein's special theory of relativity was increased by the observations of Eddington and decreased by the experiments of D. C. Miller and his associates with the interferometer, but the truth of the theory of relativity does not rest upon statistical support.

We deal perforce with probabilities because they are always involved in the problem of interpretation. For instance, what is the character of a property which belongs to each individual considered as a member of a collection but not to any individual considered separately? What does it mean to say that three out of every hundred newly-born white male babies in the United States in 1951 will die of tuberculosis? Statistical results are special cases of data requiring their own kind of interpretation. It is not the purpose of the present section to undertake a treatise on probability but rather merely to explain the relation of probability to experiment. We shall be employing the objective theory of probability: relative frequency, with chance assumed as a property of all naturally occurring events.

The probability under discussion here is the weighing of evidence for or against an hypothesis as the result of experiments conducted in order to determine it. All experimental efforts to verify or falsify an hypothesis can be nothing more than statistical, they can never be absolute (see Chapter IV, 2, above). There are degrees of truth, and that is what is meant by probability. The probability of an hypothesis, however the hypothesis itself be stated:

absolutely or statistically, is the degree of its verification (or falsification). The truth of the laws of science is not statistically expressed. Some laws are statistically formulated, but this is not the same thing. The tentativeness is built into the equations but is not part of their assertion. To say that the probability of throwing heads is 1/2 does not mean that such an assertion is half true; it means rather that it is asserted with great certainty that the probability is 1/2.

We shall have to distinguish here between empirical and mathematical probability. Empirical probability is posterior to the events estimated, mathematical probability is prior; empirical probability is concerned with the degree of confirmation of hypotheses, mathematical probability is concerned with the prediction of events. Von Wright calls the former the frequency view and the latter the possibility view. We shall treat of empirical probability here and of mathematical probability in Chapter VII. They have in common that the existence of a truth to be found is assumed in both cases: either the supposed parameter conditioning the empirical probabilities or the theorems derived from the assumed axioms in mathematical probabilities. It is also helpful at this point to recall the distinction between induction and probability set forth in Chapter III above (Sec. 2(a)), where it was explained that probability does not discover inductions but only verifies or falsifies them. Both types of probability are involved in the testing of hypotheses, but mathematical probability belongs to a later stage.

Perhaps this is a good place to indicate the approach to statistical inference which is assumed here. The Bayesian view of probability as a measure of belief is rejected as subjective and the Neyman-Pearson method of testing statistical hypotheses is rejected as irrelevant, since we are concerned with hypotheses on the assumption that they are objective and general. R. A. Fisher comes closer to stating the correct use of statistics in the scientific method so far as empirical probability is concerned.

Empirical probability has to do with the matter of proportions. It concerns the fitness between the hypothesis and the data derived from the experiments which are proposed to bear upon it, based on the range of the field which is in this way sampled, and is thus a calculation of evidence. In Peirce's conception, the ratio of the number of values which satisfy both the premises and the conclusion of an argument to the number which satisfy the premises alone. Empirical probability may be defined as the relative frequency with which the relevant formal hypothesis is supported by the favorable cases in a selected class. It is the limiting value of the relative frequency of assigned properties in a given population. The objective statistical character of the results of experiments enables them to be of service in the laboratory, thanks to the manipulation of statistics produced by actual procedures there or in the field.

Empirical probability, then, issues from the observations of relative frequencies. How often does an expected event occur and how often does it fail to occur, $P = f/f + u$, where f stands for the favorable cases and u for the unfavorable cases. A gathering of statistics is one kind of experiment, having its own peculiar rules. The investigator must make sure that what he is collecting is a fair sample, and to make a fair sample he must know something of the population with which he is dealing and he must select his instances with the proper degree of randomness. The items selected must be typical: they must have the same properties and have them in the same proportion as the whole population. If he has satisfied himself that he is dealing with a hypothetically infinite population, he must make sure that the results are such as would obtain if the samplings had been repeated an indefinite number of times. He must endeavor in short to see to it that his finite sample approximates the same proportion of different individuals as the whole class. Samplings must be repeated in order to take care of sampling variation. The type of sampling adopted, as for instance whether some mechanical method is employed, must depend upon the kind of data being studied: different types are adapted to different methods.

The degree of errors in statistical studies of this sort are important. Is the distribution of errors a normal one or not? Has the investigator taken into his account the finite mean and the standard deviation? In particular, has he avoided the error of replication whereby the limit of increased accuracy through the repetition of samplings is exceeded? There are more pitfalls along the way. If an investigator knew the initial position and velocity of every molecule in a gas and the differential equations governing each change, with each molecule considered a rigid sphere and the motions calculated by the laws of collisions in classical mechanics, he still could not calculate a given position and velocity, for such information would not rule out chance variation. Thus probability would still require us to appeal to the laws of large numbers in such cases.

The interpretation of statistical data begins with the tabulation of results. Masses of data can then be represented by qualitative or quantitative graphs. It is often possible to take one more step to make the data more easily manipulatable. Equations not derived theoretically but obtained from the quantitative graphs can serve as summations of experimental data. The ideal equation is one that represents the data most closely and requires the fewest arbitrary constants. Such equations are arrived at by plotting the data representing them approximately by means of a curve and then by reading off the corresponding equation, making the best possible guess as to its form. The constants involved may represent the operation of an instrument or they may represent a condition in nature; in either case, they need to be evaluated.

Statistical design is of course important in the strategy of research. Various statistical methods have been devised for manipulating the data in such a way as to bring out their significance for the hypothesis in question. The investigator wishes to know to what extent the data verifies (or falsifies) the hypothesis. He may have to rely on differentiation or integration, on the method of least squares, and the χ^2 square method of testing goodness of fit or on various other methods of studying distributions: binomial, Poisson, etc. He may have to introduce various types of mathematical analysis, such as the Fourrier series or periodic functions. He may have to study correlations and co-variance.

The results of interpreting statistical data consist in what are called statistical laws. Statistical laws are predictions that the future will resemble the past in at least those respects which rendered the particular statistical data in question possible. Statistical laws are never absolute, statistical formulations never completely determined. But they do tell us something about the average behavior to be found in that segment of nature which has been under examination. The high probability that, say, 85% of the *A*s have been *B*s should not be taken to mean that the next *A* will be a *B* but rather that the proportion of *A*s which are *B*s will remain the same in all future samples.

After the adoption of hypotheses, the scientific method becomes a method of constructing evidence. At the beginning this may consist in the intersection of a number of different kinds of experiments; but there are stages beyond experiment, and these introduce mathematical calculations into the construction. The hypothesis may now be framed in the mathematical language, and now looks out to other propositions at the same level of abstraction.

Science proceeds on the assumption that all facts are facts in a system; but until we are thoroughly familiar with a sufficiently wide system, it will continue to be easier to reach the system by way of the facts than the facts by way of the system. Hence experiment must continue to be more important than calculation. The strongest argument for the existence of laws in nature is the regularity of the divergence from law, for regularity is law and there is no divergence except in another legal direction. Deduction from theory to facts cannot be relied on to reach all of the possible facts until it becomes possible to explore all deductions. There are many provocative facts which were within the reach of calculations but were never discovered until experiment made them available.

On the other hand, it should never be forgotten that although all empirical knowledge is sure to be partial and statistical, it is the absolute truth at which the scientist is aiming, and did he not suppose it to exist he would be unlikely to pursue it, and this is no less so because it lies beyond his reach forever and remains, as Jevons put it, "the distant object of his long continued and painful approximations." For the closer the approximations the greater the victory for science.

We must turn next, then, to look at the mathematical treatment whereby the empirical results obtained by the hypothetico-experimental method are systematized.

CHAPTER VI

THE TESTING OF THEORIES: CALCULATION

1. *The stage of mathematical verification*

When an hypothesis has successfully passed the stage of testing by experiment and is assumed to be to that extent verified, it is passed on to the next stage which involves further verification by mathematical formulation as a test for consistency. Little by little the scientists extract from nature some small pieces of knowledge about it and then endeavor to put the pieces together to form an intelligible picture of the whole. This is accomplished first by observation and experiment and then by the use of mathematics. Mathematical reasoning is a way of extending in the direction of formalization the work which the senses aided instruments have begun. At this second stage the hypothesis is called a theory. This means that as a formula the theory is proved if it can be deduced from other formulas already accepted or established. Generally speaking, a theory is an hypothesis for which there is demonstrable experimental evidence. Such evidence is interpreted as having furnished firm support, although never in sufficient amounts to warrant ending the investigation. A theory thus lies in strength somewhere between an hypothesis and a law.

It is fair to ask why further inquiry is pursued in the effort to justify a theory which already has been shown to square with a sample of the relevant facts. If science rests on the data disclosed by observation and experiment, what more can there be? The first answer is that science begins with the data of observation but does not end there. Science is engaged in the business of combining empirical with logical evidence. It is by seeing a proposition illustrated in practice that the investigator is led to believe that it can happen; but it is only when logic gives clearance to a happening by showing that there is nothing inconsistent involved that he is prepared to settle down with it and to admit it to the corpus of accepted scientific theories. By "empirical evidence" is meant of course the results of experiments, and by "logical evidence" mathematical formulations.

Experiment is a kind of analysis, but mathematics moves in the direction of synthesis, and the scientific value of synthesis is almost as great as that of analysis provided it is a synthesis made up of analytic elements. We have only to think of some of the notable comprehensive mathematical theories in science, such as Maxwell's equations, which united electricity and magnetism, or Einstein's theory of relativity, which subsumed both Newtonian mechanics and some of the data Newtonian mechanics could not explain: the Lorentz transformations, the Fitzgerald contractions, the Minkowski four-dimensional continuum, and the null results of the Michelson-Morley experiments. And there are many other examples, such as Dirac's work, which not only brought together the magnetic moment and the angular momentum of the electron but also gave the correct formulas for the fine structure of the hydrogen and X-ray spectra and correctly interpreted the strange Zeeman effect.

When a science fails of synthesis, this is regarded as a difficulty. Such is the case today for instance with microphysics. There are a number of particles – an enormous proliferation of types of Baryons, of Bosons and of Leptons – which have not yet been reduced to systematic neatness despite the efforts which have been expended in that direction. Again, microphysics and macrophysics have not been properly related; quantum mechanics, thermodynamics and relativity taken all together still lack the proper consistency.

At the level of calculation, a theory is no longer referred to fact. It has been got off the ground by then, and there is some hope of keeping it off. In a second test there will be an attempt to determine whether the theory is consistent with other theories previously established. This will affect intimately its more inward nature, for it is the theory as equation, which is to say as generality, that is touched by its compatibility with more firmly established neighboring equations. Just as experimentation tested the hypothesis in relation to its outward nature, so calculation tests it now, as theory, in relation to its inward nature. Intension is more to the fore now than extension. Thus it is not the relation of the theory to fact which is involved at this stage in the proceedings but its full character as a general proposition, independent of any and all of the individuals to which it applies.

The scientific requirement of increasing generalization is unremitting. It was present in the steps from observation to construction and from construction to hypothesis; it is contained in the step from hypothesis to theory. Mendeleev's discovery in 1867 of the periodic table of the elements, and

Balmer's discovery in 1885 of a formula for all the frequencies of the visible spectrum of hydrogen, were important scientific contributions in the direction of increasing order and abstraction. Planck generalized his hypothesis of the quantum of energe to all mechanical systems. Einstein moved from the special theory of relativity, with its constant of the velocity of light, to the general theory of relativity with its tensor perspective; and beyond that to a unified field theory which is more general still.

The aim of empiricism is away from the concrete – away, that is to say, from the empirical. The problem of scientific procedure may be formulated somewhat as follows. Given the results of observation, to find how far it is possible to move away from the empirical by means of abstractions and still keep them in agreement with – and answerable to – observation. Not all science is laboratory science, as more than one scientist has had occasion to remark. There are theoretical observations and speculations concerning phenomena which cannot be operated on by means of instruments in the laboratory. The work of Newton in formulating the laws of motion, and the work of Einstein in formulating the unified field theory, were of this character. Science cannot be confined to the laboratory even though it is the results of experiments performed there that eventually become further abstracted in the mathematical formulations of theoretical science.

There emerges clearly, then, the outline of a second stage in scientific method, the stage of mathematical verification. In experimental science mathematics provides the language of high generality and makes complex analysis possible. At this point the investigator leaves the domain of individuals, of concrete objects disclosed to him by sense experience and action, and enters a domain characterized by universality of language indicating the presence of an objective real world of mathematical objects and conditions. Here the investigator is dealing no longer with an hypothesis lately derived from fact but rather with a scientific theory in the form of an abstract structure. We may with some profit remember here the first stage as we saw it through the eyes of Wittgenstein in Chapter II, and then add to it a developed account of the second stage as seen through the eyes of Hilbert. The method described by Wittgenstein, very much modified, was named first-order construction. The method described by Hilbert will be the model for second-order construction.

Wittgenstein's method, it will be recalled, was to work from the observation of facts to the formulation of an hypothesis somewhat as follows. A beginning was made by naming facts. Then the names were combined into propositions, which are, so to speak, pictures of facts. These elementary propositions were next combined into complex propositions and the complex propositions were themselves combined in ways which mirror the larger segments of the actual world. This was understood to be an abstract description of the way in which science works experimentally.

Now we must turn to a more detailed explanation of the Hilbert method. The stage of mathematical confirmation is that of second-order construction, for it shows the manner in which complex or abstract propositions are related to other complex propositions. According to Hilbert, the investigator begins with undefined terms which are combined into the primitive propositions of an axiom-set. Then, by the use of selected ground-rules – specifically those of *modus ponens* or detachment and unlimited substitution – he draws theorems from the axioms. In order to provide a model for second-order construction in empirical science, it is necessary to add that the propositions of the axiom-set are chosen by induction, and that the theorems can be interpreted concretely so as to be available for application to the material world.

This, in brief, is the building of a mathematical system. Hilbert did not invent mathematics any more than Wittgenstein invented the experimental method by which hypotheses are derived from observation. What Hilbert did was to show that the manipulation of abstract deductive systems can be interpreted as games played according to certain logical rules, in a way which lends to them the necessary abstraction. For it is necessary to forget that abstractions were derived from concrete existence and that their results will be applied back to concrete existence. They may be manipulated more freely when they are considered apart.

The distinction between the stages of experiment and calculation can be illustrated by the work of two men, Faraday and Maxwell. They were contemporaries in the first half of the nineteenth century; yet Maxwell employed mathematics in his investigation of electricity and magnetism in a way in which Faraday did not. Maxwell himself recommended the reading of Faraday's researches as excellent examples of the scientific method. Science will always require the services of the concretely oriented laboratory investigator even though his work by itself is never enough, but the additional labors of the mathematically-minded empiricist are also necessary if the internal properties of a scientific theory are to be explored so as to uncover its connections with other already established laws. The disclosure of interrelations is a kind of discovery also, and established laws in science are rarely the work of a single investigator; the empirical and logical evidence not only supports the verification of an hypothesis, turning it first into a theory and next into a law, but also reveals something of its possibilities.

The task of combining the two methods consists in filling the abstract

Hilbert formalism with a Wittgenstein empirical content. All fully developed sciences employ both methods, the empirical method of Wittgenstein and the mathematical method of Hilbert, each somewhat modified and conducted in appropriate stages. The complex propositions, which are the end products of the Wittgenstein procedure (i.e. the empirical theories), are arranged in appropriate groups and found to fit abstract structures constructed in accordance with the Hilbert formula (i.e. the mathematical systems). This fitting is a precise business, and comprises the highest stage of science.

The process is a complicated one, for further inductions are required to detect the fresh set of correspondences: a group of empirical measurements are to be associated with the symbols employed in mathematical equations. Then the relations between the empirical quantities will be made evident in the solution of the equations. Scientific laws are formalized quantitative descriptions of the behavior of matter at various energy-levels. It does not matter for this purpose whether we say that the empirical generalizations resulting from observations are rearranged and restated so that they appear to be the necessary logical consequences of a set of axioms, or that the axiom-set constitutes assumptions which, however few in number, abound in deductions which have empirical consequences at some distance removed. Consider for example the rich set of relationships which are uncovered by considering the whole theory of crystal structure as an empirical interpretation of the mathematical theory of lattices. In each empirical case, a selection has to be made from among a bewildering array of mathematical systems. How was it first recognized that group theory was peculiarly suited to the formulations of quantum mechanics? How did Einstein know that the tensor calculus was exactly the mathematical tool needed for the expression of the general theory of relativity?

The formalization in the mathematical language of information obtained at the empirical level is a variety of mapping. That is to say, certain features of the empirical world are plotted on a graph or diagram, and other features, can then be read off. No map is definitive, for as information increases new maps would have to be devised on which the additional information could be properly represented. There is always more to the empirical reality than there is to the map on which it is plotted. Thus the exposure of the inadequacies of the map are its inevitable fate as science advances. Nevertheless, by showing the investigator his way round, it is the map itself that makes such advances possible. A fertile source of confusion lies in the fact that more than one map could represent the same area, although it is to be doubted whether it would be possible for two maps to do so with equal adequacy. Maps may be crude or sharp; they may contain broad outlines only or be filled with considerable detail. A well-developed science is able to produce a map with much abundance of detail. When the detail is sufficiently fine, in Wittgenstein's metaphor, the map becomes a mirror of the world.

A sufficiently detailed map may be constructed entirely in mathematical symbols, and its representation as a map somewhat hard to detect on inspection. Indeed, the application of mathematics may require a complex set of operations consisting of mathematical calculations. If the Wittgenstein procedure is a sort of intensified construction, then the Hilbert procedure is a sort of intensified deduction, employing the technique of combining deductive steps. Calculation is shorthand deduction, involving the accomplishment in a brief space of a large number of deductive steps. It renders elaborate sequences of deduction automatic, and in the case of many calculations carried out by computers, it even makes them possible.

2. *The requirements of a good scientific theory*

An empirical theory to be amenable to mathematical treatment must exhibit a definite number of characteristics. Many a sequence of logical steps in the scientific method has been ended unsuccessfully when the theory involved revealed a discrepancy with observations, or when it was shown to be either self-contradictory, too complex, or insufficiently abstracted. It should have a certain measure of (a) agreement with observed data, (b) self-consistency, (c) simplicity, and (d) generality.

(a) A good scientific theory should be within allowable limits in agreement with the data developed by the relevant experiments. These limits will be such as are defined by statistical standards, as for instance those for accidental errors and for normal frequency distributions. There are in addition the limits prescribed by the methods of investigation; every technique of observation, every instrument, imposes its own restrictions. Other limitations to agreement between theory and data are self-imposed by the very nature of the abstractions. What is involved is the degree of refinement possible under the given set of abstractions. For instance, a table top ceases to be a rectangular portion of a plane when sub-molecular distances are considered. The abstractions may include non-observables, in which case tests for agreement are severely restricted. Suppose that the theory were to be Galton's law of ancestral inheritance: that on the average each parent contributes one-fourth of the inherited traits, each grandparent one-sixteenth, each great-grandparent one sixty-fourth, etc. Then although the manifestations of each trait are observable enough, the trait as carried by the gene remains non-observable, and this remains true despite the isolation of the genetic code.

At the stage of calculation formalization has introduced new considerations; the investigator is no longer dealing with an hypothesis tested by experiment but with a theory tested by mathematics. It is a mistake to suppose that an empirical system is a purely formal one, however, for an empirical system framed in the mathematical language retains empirical elements. In mathematical physics, for example, there are to be found physical rates of change, physical constants and variables. Scientific systems differ from purely formal systems in another important respect. In the scientific systems the axioms are answerable to the data disclosed to observation and experiment and so are held to be true; whereas in purely formal systems this is not the case, and truth is a property of only those theorems which are not inconsistent with the axioms.

(b) A good scientific theory should be self-consistent. If the theory is complex it must contain no contradictions. If it did it would not be amenable to mathematical treatment and its applications would issue in conflicts. The attempt to show that the data collected can be arranged in the form of primitive propositions and logical consequences (axioms and the theorems derived from them) involves mathematical treatment. If the resultant empirical system contains contradictions then it is taken as a sign that something is wrong. In empirical systems no attempt is made to demonstrate consistency, only to accept the disclosure of an inconsistency which calls for a revision of the system.

(c) A good scientific theory should be the simplest that the data will allow. Simplicity here means having the fewest number of elements. Theories must be stated in their simplest form. But simplicity cannot be achieved at the cost of adequacy. An inadequate scientific theory might be simple enough yet fail to meet the requirements of a scientific theory, which include saving the phenomena. A theory purports to explain the observed facts, and one that does not do so would not be considered the simplest. Thus the simplest will be complex enough if it succeeds in explaining a complex set of observed facts. What appears to be simple may prove to conceal a certain measure of complexity, for instance sunlight which was found to contain many colors.

The simple is always one level of analysis higher than the complex. The quality of life is simpler than the structure of organisms. All qualities are simple but they may emerge from very complex mechanisms. The dam in a river performs a very simple function but the mechanism that makes it possible may be complex indeed. Simplicity requires that theories be comprehensive and that deductions from them follow readily enough. This does not mean, of course, that such comprehension and such deducibility is self-evident. Much time and effort may be required for assimilation through general acceptance if the theory constitutes a radical departure from established laws, as was the case for instance with Einstein's special theory of relativity or with Freud's psychoanalytic theory.

(d) A good scientific theory should be a general one. A theory is general if no concrete individual is included in it by substitution of the corresponding category. Thus when a chemist asserts that the oxidation number of a monatomic ion in an ionic substance is equal to its electric charge, he does not mean to specify a given actual volume of hydrogen, though the law may well apply to it. The theory will be such that it can be extended to an unrestricted number of cases. It will avoid excessive specificity, on the one hand, and leave open the question of whether it is unlimited, on the other. A general theory may be infinite; but in science the investigator does not go further than the evidence warrants, and there is no certifiable empirical universal. There is no way, for instance, in which we can be sure that the conditions which hold on the earth and in its neighborhood hold equally on planets in other solar systems. Thus the degree of generality required will always be something less than universality and something more than particularity, even though both must be included in the aim of every good scientific theory.

It will be noted that the criteria listed here for determining a good scientific theory differ from those set forth in an earlier chapter for a good explanation at the level of first-order construction. The criteria there were intended to determine truth by correspondence, the correspondence between a proposition ond the relevant data. The criteria here are intended to determine truth by coherence, the consistency between abstract mathematical equations. Correspondence involves a continual reference to the concrete, the individual the particular; whereas coherence is essentially abstract, and involves reference to the universal and the general. It is interesting in this connection to compare the generality of various theories in science. Some, like relativity physics, are essentially general; others, like Oparin's theory of the origin of life on earth, are essentially particular. But to the extent to which particulars of this sort are scientific they are aimed at generality; Oparin's theory, for example, could if correct be extended to account for the origin of life elsewhere.

3. *The application of mathematics from the standpoint of mathematics*

We have been considering mathematics on its own ground. Now we wish to consider the application of mathematics from the point of view of mathematics. To accomplish this we shall be obliged to undertake the perspective of logic.

We shall assume that the investigator has been given a body of empirical data; his next problem is to find the system which fits them, for in the last analysis there are no independent data. There are, it is true, data which have not yet been shown to be part of a system; but if there were data not inherently part of some system, what would be their status and how would anyone know about them, and in what language could they be expressed? It is one of the assumptions of the scientific method that all matter and all energy is related in some way. Then, again, every language is to some extent a system; the mathematical languages are formalized, and even the colloquial languages are partially formalized. Despite the fact that the discoveries of the investigator were made without the overt presence of a system, a system is still there; and in fitting the data to a system he is only finding out for himself the system to which they belong. It should be remarked parenthetically that the existence of isomorphic systems which cover the same data does not controvert the above contention. This is a point to which we shall need to return.

The investigator finds systems into which he is able to fit the findings of experimental science, but these are not arbitrary; they are limited by the data on the one hand and by the rules of logic on the other. The axioms of the system must be so chosen that the theorems are in agreement with the results of observations. In the domain governed by the data of sense experience, although greatly extended by instruments, whatever is not factual is mathematical; and at higher levels of analysis mathematics provides the only method of getting at the facts. On the one hand, it is necessary to remember that mathematical systems are never so floated on logic that they are absolutely independent at facts but only relatively so, since as we shall presently note, the elements of mathematics themselves can always be traced back to the material world. On the other hand, it is necessary to remember also that calculations are possible only by means of consistent systems, and so this is what is meant by mathematics, that the data do not assemble themselves.

Logic and facts overlap; each contains more than the area which they have in common. That this is so leads to the notion of quasi-independence and requires the two levels of exploration; the mathematical and the material or instrumental. It is the ground which they do not have in common which makes a problem for the study of their interrelations. Then again, the data do not move outside their autonomous realm; only mathematics is capable of such a maneuver. In this sense the scientific method is merely advanced applied logic. Hence the empirical is limited by the logical: bounded above by tautologies, below by contradiction; and moreover, permeated by probabilities.

It is because of the rigidity and intractability of data that the scientist clings to them over the abstract formulas. Not that the formulas are not rigid and intractable too, for they are; only, he may have the wrong formulas, but there are no wrong facts. It is well known that when a fact and a law conflict, he saves the fact. What is a fact, however, may depend upon the perspective from which it is observed; and the system in which the law exists may be so well established that in a conflict between law and fact the fact is reinterpreted. The procedure is correct whatever the reason for it, but the reasons usually given for it are incorrect. They rest on the superior authority of sense experience, which, however, would be nothing without the final support provided by systems of some sort. What does raw experience convey without interpretation? And what does interpretation signify without a language? Our barest sensations are made meaningful by invoking abstract structures which furnish the language of interpretation, and this is equally true of the data so revealed. Systems, then, are necessary elements – as necessary as the data, though facts are usually considered the more reliable elements, and laws are valid insofar as they are conformable with fact. The laws together form a system. We test the facts by making observations and we frame the laws by means of logic.

The mathematics employed in experimental science is applied mathematics. "Applied mathematics" is not easy to define, and the terms have different meanings for different areas. Work in applied mathematics is carried on largely at one or the other of two quite different levels, identified here as the research level and the level of application of mathematics. At the research level – and from the viewpoint of the pure mathematician – the mathematics employed in theoretical physics is applied mathematics, because it is mathematics employed in the study of an empirical field. At the level of applied science and technology, mathematics is also employed. The mathematics referred to in this chapter is chiefly the former, although to be sure *qua* mathematical system the mathematics employed both ways can be fundamentally the same.

The use of mathematical systems in experimental science, is, then, one kind of application of mathematics. The need arises from below. In the following diagram we can see the way in which mathematical systems are filled out with empirical content.

The elementary propositions which are generalizations about the data suggest axioms from which theorems are deduced which are then found to correspond closely with samples of similar data. This circular system is here described *ex post facto;* many problems occur along the road which have to be taken care of, and the route is by no means an easy one to traverse. It

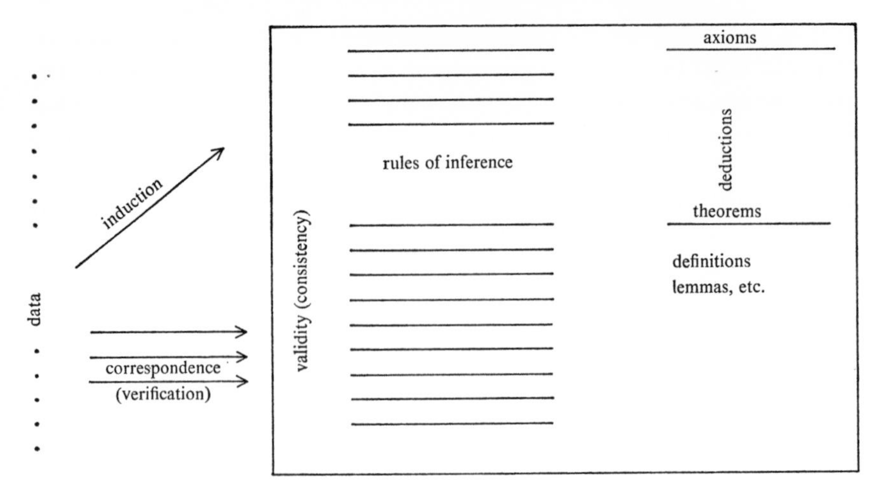

may require the services of specialists in just this combination, men who are neither pure theoretical mathematicians nor pure laboratory experimentalists.

We have before us the problem of how empirical material is meshed into a mathematical system, but our inquiry may have to begin much further back in the process. For once again after the scientist has collected his data, he is confronted with the problem of interpretation. A ghosted hypothesis guided the collection of the data yet frequently proves inadequate to cover it. Interpretation of the data is governed by the principle of economy; it must fit the least formulation required – the one just above it. Empirical science upon arriving at a properly abstract stage finds itself with a set of equations which are symbolic descriptions of its subject-matter as a result of instrumental investigations.

It is the task of mathematics, or, more properly, of applied mathematics, to help it in sorting out these propositions in order to ascertain which ones follow from the others. The success of a science is marked by the task of finding among empirical propositions those which stand in the relation to others as axioms to theorems, so that further theorems deduced from them will accord closely with observations. The decisions concerning the application of the axiomatic method to empirical material is not made exclusively by either the mathematical or the empirical requirements but consists in a nice adjustment of both. Since the explanatory value of an empirical system is its most important property, those empirical propositions which appear to be the most general laws are chosen as axioms. However, the mathematical requirements have to give way to empirical considerations in selecting those theorems which shall be the most productive.

A mathematical system when employed in connection with empirical data may be considered as a model; not a physical or mechanical model but a mathematical model. A mathematical system is an abstract model, and there is a sense in which every empirically interpreted mathematical system is a model. For instance, the Bohr model of the atom, the shell model of the atomic nucleus, the Turing theory of computing automata.

A mathematical system can of course never be verified in the same way that an empirical law can be. It becomes a model when it is found to fit some particular portion of the material world. It may happen that in it are some theorems which lend themselves to experimental testing, but this can never be more than sampling, although whatever the system fits it illuminates.

But the very abstract nature of mathematics that renders it so helpful in interpreting the data also delimits it from any kind of full description. It can be matched against the data which result from our observations only at some points; checking can never be more than a matter of sampling, and the probability of the verification of the model can never equal 1. The usual reason for this failure of verification to be no more than an approximation is a quantitative one: absolute verification would of necessity involve an exhaustion of instances, and we can never project ourselves backward in time throughout all the population of past instances any more than we could for a similar reason project ourselves forward into the indefinite future. We are limited to a few samples of the population of instances which are to be found here and now.

But there is another reason for the failure of the absolute verification of model interpretations of a mathematical nature, and this one is qualitative. Science, it has been often affirmed, is value-free, and this has been its advantage. It has confined itself to measurements of quantity and structure – to such features of the existing universe as mathematics could tell us about; as algebra and topology could tell us about form, for instance, and analysis could tell us about movement. However, the story of quantity and structure and movement is not the whole story of substance: for there is also quality. Quality may be defined here as that which is ultimately simple. Thus far, the very freedom from values which has made the enormous advances of experimental science possible has also prescribed its limitations. There are qualities in the world whose manipulation can be accomplished by present techniques but which yet evade analysis. To say of a color, for instance, that it has the analytic properties of hue (wave-length), of intensity, and of saturation, is to offer an important piece of information about the quantitative and structural aspects of the color but nothing at all about it *as a quality*.

Mathematics is the language of empiricism, but for the reasons just recited empiricism can never be reduced to mathematics. Objects possess qualities as well as quantities and structure, while scientific descriptions are only quantitative and structural. We need to develop a technique for the qualitative analysis of qualities and for their qualitative association. What are qualities *qua* qualities like, and how do they give rise to other qualities? These are questions which we have as yet no way of approaching. And in the meanwhile we must admit that despite the enormous successes of mathematics when linked with empiricism, there are other areas of inquiry which will require a quite different technique.

The unifying principle of all the elements employed in mathematics is thier common origin in the material world. Mathematical entities may have their own status independent of the way in which they were derived, yet originally they must have been derived from the data disclosed to sense experience. It can be shown for example how even so abstruse a concept as $\sqrt{-1}$ can be derived from rotation through a right angle. The Dedekind cut shows the line origin of many irrationals. Mathematics, however abstract, has a relevance to empirical structures which makes it possible for us to formulate the latter in the language of the former.

Now since mathematical elements can be derived from the material world, it should not be too surprising that mathematical theorems can be applied back to the material world. But with these proceedings mathematics as such could hardly be less concerned. This lack of concern is necessary in order to procure the necessary degree of isolation and hence to encourage production in pure mathematics. That the techniques of application were not expressly built into mathematics makes it none the less true that systems of mathematics can be concretely interpreted. The point is that the theorems may have predictive or explanatory power and perhaps even practical usefulness, while the axioms may exist only for the purpose of deriving the theorems. If there were no pure mathematics there would be no mathematics to be applied.

We have noted that if the investigator, having some specific purpose in mind, chooses the axioms of a mathematical system which has already been abstractly articulated and fills it with empirical content, then he has what is called an *interpreted* or *applied* mathematical system. More rarely, mathematical systems may be called forth by the need for empirical interpretation. But empiricism in any case never completely controls the situation. What the investigator is doing – and this is the chief feature of advanced science – is to make an amalgam of logical and material elements. Mathematics has its own laws and they are not the laws of the empirical but of the logical domain. They enjoy an autonomy which extends beyond anything which is available to empirical science. But the mathematical laws must be filled in by the formulas made up out of the regularities which are revealed to the experimentalist both in the laboratory and in the field by the instruments which extend his sense experiences and bring him reports of a reactive nature. The reason for the axiomatization of information becomes a program for the formalization of the data of experience.

4. *The application of mathematics from the standpoint of empirical formulations*

Three related but distinctly different services are performed by mathematics for propositions which, as hypotheses, have passed the experimental tests. These are (a) to make expression and manipulation possible; (b) to make new discoveries deductively; and (c) to show consistency.

(a) Theories are constructed from hypotheses by means of the mathematical language. The importance of observation and experiment of course is not lost to sight. The investigator wishes to consider chiefly those abstractions which have been forced on him by experimental confirmation. All calculations must in the end be found to be in agreement with experiment. Theories are, after all, constructed from fact if they are scientific theories. Quantum mechanics would never have been formulated without the mathematics of wave mechanics – in particular without the Fourier series; general relativity found expression in Levi-Civita's tensor calculus; the study of crystal and molecular structure was made possible by the algebra of group theory. Every scientific theory transcends the observed facts; but this is not to say that such a theory is possible without the observed facts. Indeed it works by an interpolation of the observed facts.

On the other hand, however, there is the necessity for a certain degree of abstraction in the mathematical expression of the facts. An empirical collection of data is not amenable to scientific treatment until the proper abstractions have been formed from it. The expression and manipulation of scientific formulations depend upon the viable nature of the empirical abstractions. The observed facts must be stated abstractly if they are to receive mathematical expression. For some branch of mathematics will always be found to be the appropriate language for the expression of a particular set of empirical generalizations. Equations are ways of representing phenomena so that what characterizes them is best brought out; but there are limitations in this direction also. The phase rule of equilibria in heterogeneous systems was announced by Willard Gibbs in 1878 but lay neglected for some ten years due to its complex mathematical expression; and it was not until Roozeboom's experiments and Bancroft's restatement that its importance to chemistry was recognized and further experiments conducted in the light of it.

Scientific systems are mathematical systems empirically interpreted, that is to say, systems of pure mathematics filled with empirical content. It often happens that a branch of mathematics was devised for use in a given science; for instance differential equations and the science of physics. Off-hand it might seem as though any branch of mathematics would be applicable to any empirical field; actually, some branches are more peculiarly suited to some fields than to others. It is easy to see why this must be the case. Different branches of mathematics are expressions of different kinds of abstract structures, and so have a specificity for different kinds of empirical situations. Theoretically, one could set up any kind of abstract description and then look for some concrete situation to which it might apply; and, if the work of the imagination were to be added to those which occur outside of human nature, one could be sure of finding some concrete situation to which it must apply.

Mathematics is not a science but an extension of logic. The mathematical system is empty of concrete content; it is the bare abstraction so far as these can be formulated. And it has an exactness and a rigor of proof which is not otherwise available to experimental science. It is true, however, that experimental science always hopes to approximate it and use it successfully in the efforts at such approximations. Applied mathematics never says anything in science but enables science to say what it says. It is the ideal of science and the language of science but is not itself science. The facts collected by means of experiment are rarely sufficient for mathematical formulation. The gaps are made up by the method of application. Consider for example how the mathematical treatment of chemistry by Dalton and Berzelius enabled that field of investigation to get started as a science. Thus experimental science is rarely if ever up to the deductive method of mathematics – not, at any rate, until it has reached the advanced stage of some developments in physics and chemistry.

(b) Science progresses by discovering new facts (which then call for new theories) or by discovering new theories which call for new experiments and through these for new facts. The second kind of progress can be developed independently of the first. In this second way, mathematics enables science to make new empirical discoveries deductively. The method of deductive discovery proceeds somewhat as follows. The requisite mathematics is constructed abstractly. Then deductions are made to theorems by means of the successive applications of the rules of inference, until finally theorems are reached which can be interpreted concretely in terms of observed facts. When the results of further observations or experiments are found to bear out the expectations to which the concretely interpreted theorems have led, the discoveries are said to have been made.

Examples abound. It was by suggestions from the equations of electricity that Maxwell was able to understand how light might be an electromagnetic phenomenon. From the hypothesis that electromagnetic phenomena are periodic he went on to conclude that electromagnetic waves must exist, and then showed that visible light consists of electromagnetic waves of given frequencies. His brilliant guesses made possible the further prediction of electromagnetic waves with finite velocities approaching the speed of light, thus anticipating the measurements made by Michelson and Morley. The most famous instance perhaps is the discovery of the planet Neptune by Adams and Leverrier working independently of each other in 1846. The perturbations from its orbit of the planet Uranus were accounted for theoretically on the assumption that a body of a certain mass and period could if it existed explain the discrepancy. Accordingly, the German astronomer, Galle, found the planet at the predicted position. This amounted incidentally also to a confirmation of Newtonian gravitation. Another good example is the extrapolation to atomic and subatomic phenomena of Maxwell's electromagnetic equations by H. A. Lorentz, who was led by the suggestion of the discontinuous nature of electricity to posit the unit of the electron. Still another good example is Einstein's theory that the photo-electric effect must lead to the restoration of the corpuscular structure of light, and this in the very teeth of the explanatory successes of wave mechanics. A last example: Heisenberg used the theory of matrices to discover the stationary states of subatomic particles.

It is difficult to think of mathematics as a tool of discovery, and yet it is. At deep levels of analysis, the degree of conceptualization necessary for grasping the nature of the entities and processes at work may be such as to require mathematical expression as a precondition. This is the case with quantum mechanics, for instance. The commonplace behavior of matter at ordinary levels is often the effect of causes so complicated that no amount of observation and experiment would serve to reveal them, and the assistance of mathematical theories is essential to their true understanding.

(c) When an investigator thinks of mathematics in connection with experimental science, it is chiefly in terms of consistency that he thinks. Roughly speaking, at this stage it is laws rather than facts that are employed for purposes of verification. The procedure is designed to avoid contradiction with theory for any hypothesis. Experimental science discovers certain relations. Then by means of mathematics a place is sought for these relations in a more general scheme. Such applications of mathematics usually go only so far as it is necessary to go in order to put the data together. Experimental laws in physics, for example, consist in functional relations between variables. Such functions are expressed as equations, and we get not single isolated equations but systems of equations, such as Maxwell's equations for electromagnetics. Science is at this point at the farthest remove from empiricism, since data which do not yield to some mathematical system are suspected of perhaps not having been correctly derived from facts.

It is here that the properties of logic encounter the conditions of matter and uncover obstacles. Mathematical systems are completely formalizable, empirical systems are not. And so to frame an empirical system in a mathematical language always reveals inherent shortcomings. There are obvious limits to the consistency allowable when dealing with empirical material. A system in experimental science is never more than a partially formalized system. Terms described rather than defined, and explanation of what it means to be an axiom in the system, rules of inference and sample interpretations – these are all that can be allowed and still preserve the kind of freedom which will permit the system to grow and develop as an open system. For the laws of science are never absolutely fulfilled, just as they are never entirely abrogated. There is always in any concrete situation a surd element, and there are always extenuating circumstances.

In any material system there exists an inverse correlation between degree of formalization and degree of completeness: the more the formalization the less the completeness. Partially formalized systems include more than do rigorous deductive systems, more distinguishable elements, more degrees of freedom. A gas is a system with as many degrees of freedom (i.e. independent coordinates) as molecules. But such a system can be described only in terms of relative frequencies. A highly inclusive system is usually a statistical system.

Despite the pressure of inherent disorder, of chance and of accident, the direction of abstraction toward order remains in empirical science. In addition to the analytic direction of scientific research, a direction in which mathematics can be of immense assistance, there is in all advanced science a synthetic direction as well, and it is almost entirely mathematical in character. In the direction of synthesis the empirical field of a given science is represented by means of equations. These become incorporated in a system of equations as more and still more general laws are developed. The movement from the differential equations of special relativity to the expression of general relativity by means of the tensor calculus gave greater scope to the setting of boundary conditions.

Finally, it must be asserted that the mathematical expression of experimental findings can never be more than approximate, for the following two reasons. In the first place, purely logical inference is unaffected by empirical considerations. It has absolutely no effect upon the theory of statistics that this branch of mathematics has been employed in studies on population. In the second place, empirical knowledge can never be completely formulated. The exact correspondence between *a priori* assumptions and empirical observations will never be found; but neither is it necessary, for an approximation is quite sufficient. The similarity on which such an approximation rests is itself statistical.

That there is always at most one master-theory in any advanced science, and that this master-theory is never as wide as the science, contains a further paradox. The theory of relativity does not include thermodynamics. Evolution is not as wide as biology so long as the study of organisms independently of their history, as for instance in anatomy or neurophysiology, is part of biology. No mathematical theory in a science ever embraces the whole of that science, and this proposition if true illustrates the transcendence of the world over any calculations concerning it.

At this point the investigator is not dealing with isolated situations represented by equations but rather by systems of equations best described as rich relationships. Heisenberg, for instance, considers that there have been four such systems in physics: Newtonian mechanics, thermodynamics, electromagnetics, and special relativity resulting from the theories of Newton, Clausius and Carnot, Maxwell and Einstein, respectively. (Einstein, it might be remarkable parenthetically, helped to consolidate the theory of quantum mechanics which was based on the work of Planck.) A sixth is envisaged by the prospects of constructing a unified field theory, incorporating previous work in gravitation, electromagnetics and nuclear forces.

No doubt as other sciences advance in the organization of their empirical formulations equally rich systems of theories will be mathematically discovered. There is, however, no danger of being untrue to the principles of empiricism so long as any formulation adopted can be referred back to the facts at any point. This is not to say, of course, that some very abstract formulations cannot be constructed. Where would modern physics be without partial differential equations, without vectorial analysis, without noncommutative algebras, without matrix mechanics? Yet eventually such abstruse formulations, like the simplest of earlier ones, are answerable to the relevant experimental conditions.

5. *Advanced mathematical verification*

When new information acquired by means of observation and experiment is given mathematical expression, one important result is the degree of verification which is added to it by this procedure. Experiment means for an hypothesis a measure of confirmation by the truth-condition of correspondence; it means that the facts do not conflict with the hypothesis, and his lends to it a certain support. In like manner, mathematical expression means for a theory a measure of confirmation by the truth-condition of consistency. It means that the hypothesis (now called a theory) is not inconsistent with other relevant theories. The first mathematical step is, then, to give expression to the data resulting from experiment, and the second is to show the consistency of the new mathematical theory with older theories.

Let us consider first some examples of mathematical expression for experimental results. When Kepler in the seventeenth century formulated his three laws of planetary motion, he summarized and systematized a great deal of observational data accumulated by previous astronomers, and in this manner prepared the way for Newton. Ampère in 1820 framed the experiments based on the work of his predecessors in mathematical terms which actually laid the foundations for all subsequent electrodynamics. Bohr's theory of the atom, first announced in 1913 and since confirmed experimentally by Rutherford and others, has been replaced since then. But Bohr at the time was able to combine a planetary model of the atom with some of the ideas contained in quantum mechanics: essentially, that electrons can be located only in certain stationary states of energy quanta.

Now let us look at some examples of the second mathematical step. Faraday was one of the first investigators to take this step when in the nineteenth century he combined the current atomic theory with the theory of electricity. Hamilton at the same time saw the formal resemblance between optics and Newtonian mechanics, though he failed to pursue it. Toward the third quarter of the nineteenth century, as we have already noted, Maxwell brought together in one comprehensive theory the known laws of optics and electromagnetics. Einstein unified dynamics and gravitation by means of the general theory of relativity. Schrödinger was able to show the mathematical agreement between wave mechanics and quantum mechanics.

When a new mathematical theory explains phenomena better than the old ones and is able to retain some of the old ones, it marks a definite advance in the science. The process at this stage consists in inductions from theories (now called laws or systems) to laws or systems still more general. This is surely what happened in the case of relativity theory, in which Newtonian mechanics has been preserved, some of the data which lay outside the Newtonian system has been explained, and many new experimental and observational opportunities were provided. There have been attempts, only partially successful, to bring together quantum mechanics and relativity mechanics. It has been shown, for instance, that the mass and the velocity of the electron vary in a way which is consistent with the theory of relativity. Sommerfeld had some success in explaining the fine structure of spectral lines in quantum mechanics by means of relativity. But much work yet remains to be done to bring the two theories together.

Information derived from the data of experience does not have to remain at the level of experience. Scientific knowledge issues from concrete fact but consists in general propositions of an abstract character. It is just here with the stage of advanced mathematical confirmation of theory that the degree of empiricism is lowered and the conceptual scheme taken as an independent set of abstractions without regard to its origin or development in observation, construction or experiment. The theoretical component of a scientific formula varies with the degree of empiricism, and while empiricism cannot be and indeed should not be altogether eliminated, the aim of scientific method is in the direction of the lowering of empiricism and of the raising of the theoretical component. So long as the investigator must count upon the prediction and control of phenomena to give the final confirmation, no scientific formulation can ever be empiricism-free. However, that is not the point, for the scientist would not wish it to be free in this way; the point is that a formulation which has been correctly derived according to the logical structure of research will achieve the character of high abstraction as a matter of procedure, and that, further, the new status will remain unchanged by the retention of a continued relevance to concrete fact.

6. *Difficulties of final formulation*

We have already taken notice in Section 4, above, that there are limits to the completeness of a mathematical system and that this is no less true when the system is applied to an empirical subject-matter. Godel's theorem has offered the proof in the case of a mathematical system. Systems can never be complete, and this indicates limits; but there are advantages to such limits, for they serve as warnings that all systems must be kept open. The scientific method, as we have noted again and again throughout this work, is a self-corrective method and so it must be maintained in operation. Repetitions of the scientific method are possible only so long as system construction is incomplete.

Problems arising in the course of the development of theoretical systems are decided either by logic or by fact. The structure of the system is provided by logic and the contents suggested by fact. But then there are other, more difficult cases, such as the existence of incompatible theories. In such a case, either one of the theories must be discarded, or both. If logic is not able to decide (and it cannot when there are serious inconsistencies), then the decision is up to fact. But it often happens that there are seemingly inconsistent facts, the wave- and particle-phenomena of light, for example. Systems at a high theoretical level in a science are explanatory systems and not so easy to dismiss. The powder model of the atom has been adopted without altogether giving up the Bohr planetary model, for instance. A system once established remains in force until another is sufficiently prepared to take its place. The Copernican system of astronomy is still used in some quarters.

Inevitably difficulties do arise, however, and systems are proved unsatisfactory, either because of imperfections which develop within them or insufficiencies which show themselves from without. A number of different kinds of difficulties may appear which had not been evident in a theory when it was first framed. Data adduced in support may have had to be reinterpreted so that with the new relevance they no longer carry the weight they once did; or fresh data may be discovered which do not offer support; or modifications in the expression of the theory may reveal inconsistencies which before had remained hidden. All systems sooner or later must be replaced, supplanted, or at least subsumed under wider systems; in the meanwhile, of course, despite their usefulness and authority they are properly to be regarded as provisional. The drive toward more comprehensive theories is aided as much by the difficulties which are uncovered in existing theories as by the insistence of facts they have failed to include.

The frontiers of science are marked by irregularity of advance. New facts are being uncovered which existing theories do not include, new theories are being advanced which existing experiments do not test. There is of necessity some confusion, much disagreement and considerable change. Any science of appreciable size of achievement needs the services of two kinds of specialists. It needs the discoverer without whose services the science would never have existed in the first place; but it needs also those thinkers whose work of consolidation makes order possible and with it the learning process by which new men are introduced to the existing body of knowledge in the science. The accumulation and arrangement of gains is no less a contribution to science even thought it be of an ancillary kind. If science looks to mathematics for its expression, it looks in the same direction for the criterion of consistency which alone can determine whether a scientific advance can be made an integral part of its discipline.

7. *The aim of deductive structures*

We are chiefly concerned in this work not with mathematics as such but only with mathematics so far as applied mathematics can be considered an interpretation of nature. We have noted that mathematical systems as employed in experimental science furnish the demonstrations of consistency. Newton acknowledged causality in his first two rules of reasoning. Planck confirmed it in quantum theory despite objections from Bohr and others, and scientists like Born and Waddington have reasserted it for all of science.

The interdependence of fact and theory is the salient principle which emerges from a study of actual laboratory procedures. When Eddington observed, somewhat sardonically, that he could never believe in a fact which had not been confirmed by theory, he was pleading for one side of a question which has always been argued more eloquently from the other side. For the empiricist has the prejudice that to the contrary it is the theories which have to be confirmed by fact.

Actually, both are true, although when faced with a conflict between fact and theory, he is obliged to save the facts. Many an experimenter from Fresnel on has thought that he could devise a crucial experiment to decide between the particle theory of light championed by Newton and the wave theory of Huyghens; but the rival theories remain because of the irrefrangible nature of the facts in both cases: the mass, velocity and momentum of the particles, and the wave-length, frequency and amplitude of the waves. But the facts thus saved are themselves presumed to be included under some single theory, even though a satisfactory one is perhaps not yet known, for the facts are meaningless taken by themselves. Whitehead was fond of pointing out that there is simply no such thing as an isolated fact considered in itself and apart from its place as an element in some system.

And where does the criterion of systematic consistency lead? It leads to greater generalization and greater abstraction. The tendency in empirical science at the mathematical stage is to combine laws under still more comprehensive laws. This can be done deductively, as when Boyle's Law, Gay-Lussac's Law and Avogadro's Law were all shown to be deducible from the perfectgas equation; or it can be done inductively, as when Newton inferred the existence of a law of gravitation which could include as special cases the earlier calculations of Copernicus and Kepler. It happens that without the investigator intending it the mathematical systems by means of which the experimental findings can be formulated do more than demonstrate consistency. For their limits are the limits of completeness. Investigators as far apart as Peirce (1896) and Heisenberg (1952) have dreamed of an ideal of explanation in which the conditions adequate for the description of the entire world of nature could be deduced from one grand major premise or expressed in a single equation.

Such a goal is perhaps impossible of attainment but still constitutes a limiting case toward which the increasingly inclusive formulations of the more developed sciences tend. There are limits to the mathematical method as prescribed by empiricism, however, and they show up most forcefully in the social domain of investigation. The behavioral sciences have to cope with the fact that in some areas partially formalized systems may exhibit greater power than fully formalized systems; – in the analysis of the colloquial languages, for example. Then, too, as we have already noted in another connection, premature mathematical treatment for concepts not properly abstracted may be stultifying.

The effort to formalize empirical findings is limited by the brute fact that the empirical world lacks the precision of logic. Thus while it is true that logic never applies absolutely to material situations, that is to say, with logical rigor, still no empirical proposition is both true and self-contradictory. Mathematical formulations are ideal; empirical conditions are at best approximations to rigid formulas. The world of concrete objects in motion in time and space lacks complete logical rigor; everything flows into everything else, and attempts to make final demarcations are sure to prove only limiting cases. Thus the formalization stage of verification in the scientific method is never complete. It is often assumed that such incompleteness in scientific research is imposed on it by the fallible human effort, and this is often true. But it is not entirely true, and the human element is not entirely responsible; it is just as much the objects of science – the material objects – as the subjects, which are to blame. For both the subject doing the investigating and the objects investigated are concrete and particular, and so possess equally the approximative limitations of the concrete world in general.

Suffice to say that the bonds between mathematical calculations and the results of instrumentally-aided observations and constructions are being drawn tighter as science progresses. If material truth be defined as agreement with observed fact and formal truth as tautology, then the scientific method as a whole is the effort to ascertain how closely the agreement with observed fact can approximate to tautology in any given case. If an image may be permitted here, it is that of the scientist as a climber, ascending the ladder from empirical data to a lofty position among the abstract deductive structures of mathematics whose elements though having their origin in the actual world now ride independently above it. There, from his perch he is able to study the systems and to obtain from them, as from a large-scale map or blueprint, an increased knowledge of the world of facts spread out far below him, as he takes advantage of the now unobstructed view of its expanse. This is his dream; however, in his waking moments he conducts himself professionally more like a tightrope walker who must maintain a precarious balance between an intractable finitism with its discontinuity as imposed by empiricism, and a logical perfectionism as demanded by mathematics. Each has peculiar characteristics which forever preclude it from being assimilated to the other. And yet we can explore one only in terms of the other.

8. *Mathematical probability and causal law*

We have already discussed empirical probability in Chapter V. There we saw that empirical probability meant the verification of hypotheses, whereas now we shall see that mathematical probability is concerned with the prediction of events. Mathematical probability is a different kind of probability; it is purely theoretical, and consists in predictions of the relative frequency of occurrences as these can be deduced from axioms and theorems. It is usually conceived to be the result of laws modified by chance. Empirical probability is *a posteriori*, mathematical probability *a priori*. Empirical probability was concerned with experience, mathematical probability is concerned with expectancy. Mathematical probability issues not from observations of relative frequencies but from calculations of predicted frequencies. It is derived not from statistical evidence but from chances estimated beforehand by means of established formal procedures.

The distinction between the two types of probability has an important meaning so far as the verification of theories is concerned. Empirical probabilities are calculations made from the results of experiments. Mathematical probabilities result from calculations of mathematical predictions and so establish the type, and influence the design, of experiments which it is necessary to carry out. Agreement of experiment with calculations means here that the mathematical probability calculates the expected results, and then experiments involving empirical probability (i.e. statistics) agree or disagree significantly with the calculations.

Calculation is thus used as a kind of hypothesis to guide observation. The experiments and observations employed here are at a far more complex and sophisticated level than those at the starting stages of scientific procedure, as discussed in the earlier chapters of this book. Here they are involved in probability formulations and their underlying causality constitutes the last part of the test of law, a law which is no less a theory for having been before that an hypothesis.

The claim has been made for the absoluteness of probability also, for instance by von Mises. For him there is no such thing as causality, only the averages of statistical behavior. The point of view of von Mises and of those who follow him represents the same kind of extreme view as the one claimed for causality by Laplace. The argument against von Mises is somewhat as follows. If there were no causality in the old sense, only statistical averages of random behavior, then all would be chaos. But this is impossible, as Peirce (1879) and Bohm (1957) have shown, for chaos is an unworkable ideal: averages of statistical behavior generate their own order. Given a sufficient number of individuals of any kind, there will begin to manifest itself a certain similarity in the behavior of some of them, and this tendency to order will grow. Complete chaos is as impossible in the actual world as complete order. But the tendency to generate order in statistical averages is as strong as the opposite tendency – which also exists – of order to fall off into chance disorder. Chance prolonged produces uniformities. Random events allowed to fluctuate freely arrange themselves according to statistical laws, which is the beginning of order. A good example is proposed in Urey's and Miller's primeval "soup" as the source of the origin of life on the earth. In short, causal laws are what empirical generalizations approach; that they tend to fall off somewhat does not alter them but only limits them.

The hypothesis of an order in nature cannot be reduced to the hypothesis that this is the way in which we read regularities. For the latter implies a distinction which the former need not admit: the order *of* nature is also the order *in* nature, and in reading regularities we are simply recognizing the existence of various orders. As Meyerson (1920) has argued, in causality, despite the prevalence of change, regularities are interpreted as exhibiting the existence of identities persisting through time. Causality is a concrete exemplification of both *modus ponens* and unlimited substitution, so that the rules of inference come out of the causal situation as abstractions from it, for the displacement which is demonstrated is that antecedent cause is *substantially* identical with the consequent effect.

Thus representing natural laws as functional relations between variables expressed as equations, on the basis of a few experimental samples, is to suppose a hypothetically infinite population of which the equation is always true, or at least always as true as extenuating circumstances will allow it to be, which is never altogether true and always partaking somewhat of truth.

On the other hand, it is not possible to eliminate probability in favor of causality, either, because causal mechanics are limited affairs and also because they cannot be completely isolated. If Bohm (1957) is right when he says that the infinite richness of nature prevents mechanisms from ever being universe-wide but keeps them always as causal descriptions of prescribed domains, then not all statistical situations can be reduced to causal laws, and the existence of parameters behind statistics does not mean that the parameters can be reached by the statistics. The chance fluctuations are brought about by the fact that no system of phenomena under investigation can ever be completely isolated, not because the investigator lacks the proper abstractions to describe the situation but rather because the isolation itself is never completely effective. No laboratory or field condition can ever be arranged so that all contacts and relationships outside the system to be isolated can be altogether avoided; no provision can ever be made that will successfully eliminate all contingencies.

Both causality and chance, then, are limited, and each is limited by the other. Whether we say mathematical probability and predict the average behavior of large numbers of events, or deterministic causal law which has to be modified by the extenuating circumstances upon every application, is a matter of language. We are changing the system of concepts but describing the same set of conditions. For the conditions *can* be looked at from different perspectives, and a different perspective is what a different system of concepts essentially is.

Existence contains both elements as ingredients. Every event is partly caused and partly the result of chance. In a perfectly isolated system, cause alone would operate; but such a degree of isolation is impossible of attainment. And so we say that if A causes B, then given A, B will happen, all other conditions being the same. It is not possible to eliminate causality in favor of probability altogether so long as mechanisms do exist. Good examples of causal laws are: the laws of motion of classical mechanics, the second law of thermodynamics, and all the conservation laws. The second law of thermodynamics has never been successfully challenged even though evolution seems to work in the other direction. And there are chemical reactions which can be reproduced in the laboratory; the electrolysis of water, for instance. Many of the particular chemical combinations by which two or more elements compose a compound are known and are invariant. Even at the level of quantum mechanics, the known reactions are constant; that is to say, they can be brought about by the proper experiments. The laws of gases operate inflexibly despite the fact that we do not know which molecules in particular will produce the average effects. And the conservation laws in physics: the conservation of mass, of energy and of momentum, are reliable causal laws.

Thus it appears that causality and probability must be regarded as different types of interpretation in smaller or larger domains, where neither is absolute and where neither covers the whole of nature. The laws of science are not laws which phenomena must "obey" as though they were dictates from on high, on the analogy of laws established by governments and enforced on individuals. Neither are they the statistical averages of individual behavior having no causal efficacy. No; the laws of science are elicitations of the regularities which exist in phenomena; how and why they behave as they do are as one the same: they behave in such and such a way because that is the way they are. Thus the laws of science are the phenomena of nature asserting their own irrefrangibility. The attitude toward them ought always to be based on the probability of their holding up under the discovery of new and challenging facts in the future. In short, causal laws are absolute laws but their acceptance must always be a matter of probability. This is the only sense in which probability is subjective: we never absolutely accept the total verification of causal laws.

If, then, we accept the validity in their own domains of both causality and of probability, we still have some thorny problems ahead. What, for instance, is the relation between them? Again, causal laws differ from statistical laws in being extreme cases, for they are statements of *complete* generality. But complete generality requires complete abstractness, and we have just noted that empirical laws can never be completely abstract. How, then, is this difficulty to be resolved? In order to discover this we shall have to look a little closer into the structure of causality.

Causality may be defined as the necessary action of one material object on another. A causal law is a universal proposition having an empirical content. From the formal aspect of the proposition it involves logical necessity: e.g. "All *X*'s are *Y*'s"; while from the material aspect it is an extreme case of statistical probability: "100% of the *X*'s are *Y*'s. According to E. H. Hutten (1956), empirical causal laws are counter-factual conditional propositions, expressed by sentences in the subjunctive mood. "If there is no interference, then a force exerted between any two particles varies directly as the product of their masses and inversely as the square of their distance." But there always is some interference, e.g. the atmosphere; thus the antecedent is never true. When the law is stated for a vacuum the facts are idealized, and the law then becomes a simple deduction. But we are closer to empiricism when we leave the facts as they are and state the law ideally so as to allow for its universal application modified by extenuating circumstances.

Two contributing theories will be helpful here. One is Cournot's theory of the existence of separate causal lines (1861). The other is Bohm's theory that causality is not always a one-one relationship.

Cournot began with the idea that there are causal sequences, though not every occurrence belongs to such a sequence. He then went on to the analysis of isolated causal sequences as strings of events so related that something can be inferred from one about the others regardless of what is or is not occurring outside the sequence.

A statistical law is an average of intersections. One might argue, then, that there are as many degrees of freedom as there are causal lines. The effects of separate causal lines could offer mutual interference, and this would be a chance event. A concourse of causes acting independently of one another might bring different forces to bear upon a single isolated fact; each time the causes met they might be differently sorted and have different weights, so that despite the presence of causality, their effects could only be averagely predicted. Cournot's theory of separate causal lines is very reminiscent of Whitehead's "historic route of occasions," and was followed by Russell.

We need now to add to Cournot a second theory introduced by De Broglie (1927) and later elaborated by Bohm (1952). De Broglie postulated that any solution of the wave equations would have to be doubled by the addition of a moving singularity to the interpretation of the wave phase, and Bohm supposed that causality and probability are related by means of the integrative levels, causality always being one analytical level lower than its chance effects. Thus for example he proposed to regard the statistics that obtain at the quantum mechanical level as effects, and to seek for their causes in the hidden parameters lying at a hypothetical sub-quantum mechanical level. Bohm, a few years later (1957), argued that absolute causality would be a one-to-one relationship with one cause resulting in one effect.

Such a simple situation is rarely found; what we are usually dealing with are one-many or many-one relationships where the effects are never determined uniquely. Mill pointed out that there can be and frequently are a plurality of causes, and Bohm has added that there can be also a plurality of effects. Small causes may have very large effects, and very small effects may be the result of very large causes. Death may be caused in a number of ways and be the same death; a mouse killed by a nuclear explosion is no more dead than one killed in a mouse trap. Causal relationships of the one-one variety occupy the same ideal status as do perfectly radiating black bodies, frictionless engines or perfect gases. If we now bring together Cournot's and Russell's theory with Bohm's, we have the situation that separate causal lines are sequences of events which are woven in relationships which determine other events causally but do not determine them uniquely. If we combine the theories of Cournot and Bohm, we can then suppose that chance is the average of effects which result from the intersection of separate causal lines lying at least one analytical level lower. In this way, the apparent conflict between causal law and statistical law become resolved. Both are true readings and neither invalidates the other.

Every scientific law, then, is both causally determined and statistically indeterminate, depending upon the levels. Causality for a given set of phenomena is always one analytical level lower than probability. The intersection of chance events always occurs one analytical level higher than the events which constitute the causes. Chance at the macrocosmic level is the result of causes lying below it at the microcosmic, and probabilities at the micrososmic level studied by quantum mechanics must be the result of causes lying at a sub-quantum mechanical level. Bohm posited this relation of levels between chance and cause for the quantum and sub-quantum level, and Bohm's law is here generalized so that it can be extended to all the empirical levels. Examples abound. Superconductivity and certain effects in thermodynamics may be the results of causes at the quantum mechanical level, for instance, and significant averages of social behavior studied by social psychology may be the result of causes lying at the biological level of neurophysiology.

The laws of sciences are abstractly stated. They have risen above empiricism, and this is no less the case because they have done so by its aid. For it is true that in science they remain forever answerable to empiricism. Observation and experiment were the means by which the hypothesis was raised to the status of theory and theory to the status of law; and when the use of mathematics enables the scientist to complicate the schematism, it is only before he refers the law to empiricism again by means of the prediction of events and the control over phenomena. To the consideration of these last two stages we must next turn our attention.

CHAPTER VII

THE TESTING OF LAWS: PREDICTION AND CONTROL

According to the logic of scientific investigation, observation leads to induction and hypothesis. The hypothesis is then tested in two ways: by correspondence with the relevant data by means of experiment, and by consistency with the existing systems in the science by means of mathematics. In the second of these two tests, the hypothesis is called a theory. It is next ready for the third and last test, which is by means of the prediction and control of phenomena. In this last test it is called a law. In other words, we have said that an hypothesis is a proposition suspected of being true but for which there is as yet no support beyond observation; and that a theory is an hypothesis which has received a measure of experimental confirmation. Now we shall assert that a law is a theory which has received mathematical confirmation. One of the great values of laws at this stage in the proceedings is that they point to further observational tests.

The final test of a law is once again an empirical one, though different from the kind of empirical test which as an hypothesis it encountered at the very outset. For the early empirical test was an experiment, while the later one is an application. One way of offering evidence for a scientific law is to give a concrete interpretation of it. In this way, applied science finds its theoretical justification in its value for pure science as constituting the last stage in the scientific method of investigation. But it is also true that applied science for quite other reasons receives a considerable amount of attention on its own.

Both the prediction of phenomena and the control over phenomena constitute varieties of application. Speaking generally, predictions have to do with advances in pure scientific theory, while control is more concerned with practicality. We shall look at prediction first.

1. *Prediction*

(a) *Predictions of laws.* Prediction in scientific method means the theoretical anticipation of natural occurrences or of experimental results. The investigator deduces from the law what events he can expect to occur or by appropriate methods can bring about, and the occurrence of those events constitutes one more confirmation of the law. That two bodies attract each other directly as the product of their masses and inversely as the square of their distance apart, can be held to be fulfilled as a prediction by the coming together of every two bodies subsequent to its promulgation as a law. De Broglie predicted the diffraction of electrons in 1924, and in 1927 Davisson and Germer performed the experiment confirming the prediction with some deliberateness, thus making the interpretation of the data in this instance a relatively easy affair. In the example discussed earlier (chapter V, Section 3) Yukawa in 1935 predicted the existence of a particle half way in weight between the electron and the proton, and it was not at all an easy feat to recognize in the cosmic rays three years later that the unknown particles had the required mass and must therefore be the predicted particles, which were later named mesons.

The original hypothesis, adopted with the aid of observation and induction, is now held to have fully matured. Its inward abstract nature has so to speak triumphed, and it is as established as any scientific formulation ever is; that is to say, it is accepted until overthrown by contradictory data or superceded by wider formulation. It is then at this stage related back to the material world, and if this relation is again taken to be a further test of its truth, it must be remembered that this test is not regarded or carried out in the same spirit as the previous test at the stage of experiment. For then it was desperately and entirely dependent upon material particulars in an experiment whose outcome was essentially doubtful; whereas now, while it is still dependent upon them (as all generalities in empirical science always are and always must remain), the outcome is no longer equally doubtful, and the probability of expectation is overwhelmingly in favor of support.

Predictions are made from the full standpoint of the law as law, and *control* is exercised in the same fashion. Both relate the law as proposition to events as actualities; but the events logically stand to the abstract formulation in the position of *subordinates.* The law prevails so long as it is a law. And as a law its success is dependent upon its power to predict effects previously unknown. Such a power is both novel and crucial; it is striking evidence of the truth of the law and it suggests the prospect of controlling events.

Despite the probability of expectation in favor of the law being tested by means of prediction and control, despite, that is to say, the treatment accorded the law as though it were reliable and could hardly be expected to fail, the scientist must not be prejudiced in its favor if it does fail, and must accept this kind of failure as a serious fault. Care must be taken to record all instances of prediction, both those which are successful and count towards verification and those which are unsuccessful and count toward falsification, in order to weigh them properly. Strict impartiality requires that the investigator present the law, which still retains something of its character as an hypothesis even at this late stage in its testing, with the facts, and he must do so as much when they are unfavorable as when they are favorable. The new law exists in the same domain as the old, and must be able to make possible predictions which hold for the same class of phenomena with a greater degree of accuracy or a greater extension of range. In any case, repetitions of the operations are possible, and the self-corrective nature of the scientific method once again sustained.

Predictions may involve as experiments, the consequences of laws utilized to obtain the predictions of actual results in the laboratory. Having determined by drawing deductions from the proposed law what conditions could be expected, the investigator designs his experiments to bring about those conditions, and after observing the results he then compares them with the deductions from the laws; if they agree he regards the successful predictions as a certain measure of verification. If the predictions are successful the law is further verified. That is to say, it will be accepted until deductions are made from it which are contrary to fact. The successful outcome of an untried experiment is strong evidence for any law to which the experiment is relevant and from which it has been predicted.

Agreement with observation, perhaps the touchstone of the experimental scientist, does not belong to the early, observational stage of science, for then there is nothing with which observation could agree, observation being at that point its own desideratum. It belongs more properly to the experimental stage when, for the first time, there is an hypothesis to be tested. And, as we shall now see, it belongs also to the later stage when predictions are made from a law and there are further observations, this time of the accuracy with which the predictions are fulfilled. Predictions may be distinguished with regard to whether they are predictions of universal conditions or of singular events: "all men are mortal" or "this man will die."

Logically speaking a prediction is a weak deduction. If for example the law states that only one electron can be found in any energy level with the atom (Pauli's exclusion principle), then the prediction can be made that equal volumes of gases having the same temperature and pressure must contain the same number of molecules, and it can be predicted also that a litre of helium will contain the same number of molecules as a litre of oxygen when their temperature and pressure are the same. If the present state of a system is known, and the functions and rates of change are known, then within the limits the future of the system can be predicted. The failure of prediction does not mean that there are no laws in nature and that probabilities prevail, only that the laws are not known. Probabilities are important in prediction in determining whether actual events bear out predictions within significant limits, for there will usually be some variation. Predictions are usually stated in ideal terms; actual events can only approximate them. There are events which have not been predicted although scientists do not doubt that they are determined. Planck (1933) has cited certain meteorological phenomena, while Chwistek (1948) thinks of a scrap of paper tossed about in the wind.

Prediction offers evidential support for a law by the continual testing against fact which it constitutes. The difference between a scientific law and some other kind of general principle in a non-scientific field of inquiry is that the scientific law is always held accountable to fact. This condition is never abandoned so long as the law is established. A law may be tested at any time, and at each time its life as a law is at stake. To maintain its status apart from fact, i.e. as an abstract and general law, it needs to continue this accountability to fact. Laws to remain acceptable must be applicable to all relevant facts, past as well as present, future as well as past. Now it so happens that past fact becomes less accessible, not more, as time elapses; whereas future fact passes into the present and becomes more accessible. Hence prediction is a way of testing laws against future facts, which must bear out the predictions when they become present.

Meanwhile, there is a certain sense in which it is legitimate to say that law may be considered standing predictions. The assertion that the rates of diffusion of two gases are inversely proportional to the square root of their densities (Graham's law of diffusion) means that when at any time in the future gases are put together they will diffuse at the rate specified and this can be fined down if in our prediction we name particular gases. It is no more than a weak deduction to say that if a law which applies to all of its instances applies to that sub-class of them which are all of those instances which are yet to occur. For a candidate for the status of law, the twice-certified hypothesis which is our theory must be further certified by satisfying sample predictions which indicate something of its likelihood of serving as a standing prediction.

Examples of successful predictions abound in physics and chemistry. Wireless waves were predicted from Maxwell's equation in 1864, and Gibb's phase rule (1877) provided for many predictions, including the existence of tautometers. The electron theory enabled Lorentz to predict the Zeeman effect, discovered in 1896. Faraday's prediction from the law of magnetic induction that the earth would be found to act as a magnet was confirmed by experiments. The law of centripetal force which accounts for the force necessary to hold a moving body in a circular orbit has been immensely useful in predicting planetary motions. The quantum theory has been employed successfully to predict the energy of particles emitted from such radio-active elements as radium, plutonium and thorium. The solutions of the field equations of general relativity by the method of successive approximations led to the prediction that waves were to be found in the gravitational field propagated at approximately the speed of light.

Again, important predictions have been made many times from the conservation laws: the law of the conservation of energy, of mass, of electric charge and of angular momentum. Debye (1932) predicted that the diffraction of light traversing an ultra-sonic field would result from the periodic variations in density in a liquid traversed by ultra-sonic waves. Shortly after, Sears performed the necessary experiment and the prediction was borne out. This is the origin of the Debye-Sears effect: a brilliant series of diffraction images gathered by a lens at the end of the trough in which a parallel beam of light had been sent perpendicular to the path of ultra-sonic waves.

The investigator relies upon known facts and accepted laws to predict the existence or the occurrence of hitherto unknown effects. He employs the known to aid him in anticipating the unknown; prediction is one of the most important techniques in the scientific method of discovery. The more abstract the formulation in a science, the more effective the prediction. Long-range effects are made chiefly from highly abstract theories. It is obvious that in making predictions from a law, everything concerned in the validity of the law is at stake; prediction is the supreme test to which an hypothesis can be subjected, the ultimate commitment of the theory. For if the prediction were to fail, confidence in the strength of the theory would be severely shaken.

It might be helpful here to compare prediction with advanced types of explanation. For in explanation we endeavor to account for the facts by means of the theory, whereas in prediction we endeavor to account for the theory by means of the facts. Combining explanation and prediction we get the series, facts-theory-facts, or, in the terms we have been exploring, observation, hypothesis, verification. Facts can never under any circumstance be left out of the account; however, they play a decreasingly important role as theory assumes greater and greater proportions.

The prediction from laws involves the extension of concepts and the anticipation of the results of findings; a law is an expectation. The tendency of science as it advances is to move toward the predictive end of the method more heavily than the observational beginning. In a complex and advanced mathematical science, the accretion of new knowledge comes faster by means of deduction from known laws to new and predicted effects than from data accumulated under observation or by induction from experiment. Francis Bacon, whose influence is felt as late as Whewell and persists still today, was talking about the early stages of a science when he insisted that knowledge had to be inductively discovered by means of experiment. Experiment will always stand in the background and so to speak monitor all scientific endeavor, but in the later stages of science it no more has to be actively employed than law-abiding people find it necessary to undergo litigation in the law courts. In the end predictions are used and experiments retained as an occasional method of verifying predictions.

(b) *Predictions of empirical systems.* Predictions of laws are one of the last two tests in the seven stages of the scientific method. We cannot leave the topic, however, without a word about predictions of empirical systems. These are most familiar, perhaps, in physics, and especially in astronomy. Both the Ptolemaic and Copernican systems were used to predict the paths of planets. The three confirmed predictions calculated from the theory of relativity of 1915 are by now well known. These were: first, the closer prediction of the elliptical orbit of Mercury around the sun than was possible by the Newtonian system; secondly, the bending of light rays as they passed near to the sun's surface; and, thirdly, the shift in the wave-length of starlight toward the red end of the spectrum as the light moves through the star's own gravitational field. Later, plans were made to measure simultaneity by the comparison of two clocks, one on the earth and one carried in an artificial terrestrial satellite, in accordance with the requirement that the signal be independent of space and time measurements, the clocks will be atomic clocks, and the experiment will determine whether a separation in space will make a difference in time, as called for by relativity mechanics.

A famous instance of prediction from theory is the positive electron (the positron) the discovery of which was predicted by Dirac, using both special relativity and quantum mechanics. The positron was actually found in cosmic ray experiments, first by Anderson and later by Blackett and Occhialini. Yukawa's prediction of the discovery of the meson years before its existence was verified experimentally in 1948 was based on already existing knowledge.

In chemistry, Lavoisier predicted the discovery of potassium and sodium. Gallium, whose discovery in 1876 was predicted by the Mendeleev table by means of a blank in the series, when found agreed remarkably with respect to atomic weight, melting point, specific gravity, etc. Predictions of the chemical composition in the case of certain compounds has often occurred and been subsequently verified. Lately, the synthesis of proteins has been predicted and successfully executed.

If now we move from chemistry to biology, we encounter a curious phenomenon. The most comprehensive theory in biology is that of organic evolution. Yet no predictions have been made from it as to the next development in organisms, say after man. Yet if the theory of evolution is all that its adherents claim for it (and there are no serious objections to it any more), then we should expect it to be treated as established empirical systems and laws are always treated in science, namely, as a source of predictions which can be later verified or falsified.

Predictions in sciences higher in the scale of integrative levels but without mathematics have been undertaken. Toynbee's attempt to predict the next cycle in history from his theory of the structure of history is an example. His loose reference to facts, however, his practice of including them when they supported his generalizations, and of ignoring them when they did not, relinquished any support which the application of empirical criteria might have furnished. The weakening of empiricism brings about a greater degree of uncertainty in the use of empirical systems and of laws established by their means. The more complex the material structure the more incomplete our knowledge of it, and consequently the greater our inability to work from it.

(c) *Deductive discovery*. The kind of predictions made at this stage of the scientific method differs considerably from the predictions made at earlier stages. Every experiment in a sense embodies a prediction made from an hypothesis. The prediction is to the effect that if so-and-so is done, the result will verify (or falsify) the hypothesis to that extent. From the known facts of the tensile strength of water and its ascent in trees, D. T. MacDougal argued that during the day when the rapidly moving water is under its greatest tension the diameters of tree trunks should be slightly smaller than at night. Later experiments verified his predictions. In the last chapter we saw the mathematical success of Maxwell's mathematical predictions of electromagnetic phenomena. From his work showing that visible light consists of waves of certain frequencies, he went on to predict the discovery of waves of other frequencies. Two decades later Hertz produced in the laboratory waves of lower frequencies which we have come to know as radio waves. And when the work of Marconi on the technological problems which presented themselves was completed, the way was prepared for radio, radar and television.

In the last chapter, deductive discovery was raised explicitly, in connection with the mathematical stage of verification. The discussion in the present chapter differs from those earlier presentations chiefly in occurring at a later stage in the development of the method, that is, as essential to the method itself, and not, as earlier, thrown off, so to speak as an incidental illumination. In previous stages, either experiments needed to be performed or calculations to be made prior to the prediction which was embodied in deductive discovery; now the prediction can be made only from the information at hand. At the advanced stage where the hypothesis has already passed from the status of theory to that of law, predictions from abstract structures to subordinate laws which can be tested against the data of experience, or to natural events in which there has been and is to be no human interference by means of experiment, are made. The scientist in the latter case is reading off from the diagram of a theory the consequences of certain propositions which can be tested against the data disclosed by experience but which without the theory might have gone unnoticed or not had its significance read. In both cases, there is a deductive discovery; in the former case of a law, and in the latter case of an event.

The prediction of law remains at the theoretical level; the prediction of an event begins with a knowledge of the starting conditions and of the law, and forecasts the discovery of an event. It is a reliable procedure that where observations are expected, theoretical expectations are first calculated, and the amount of agreement or disagreement noted. This is especially true in astronomy, for instance, where interference with the phenomena is impossible. At advanced stages of science, the phrase "agreement with observation" generally means that a prediction which was concocted from theory was found to be borne out by subsequent events.

Examples of predictions which were deductive discoveries are not far to seek. The prediction of eclipses, which so amazed the ancients, and of comets, such as Halley's Comet, are instances where calculation led to successful prediction of events and where control was out of the question. Herschel pointed to the achievement of Fresnel, whereby from a set of abstract principles not only all the phenomena of double refraction is deducible but also facts later verified by experiment but not known before and indeed contrary to the going opinion that crystals possessing the property of double refraction undergo a deviation from the original plane. Darwin has reported that this theory of the origins of the coral reefs was executed deductively before he had seen a coral reef, and indeed then it was necessary only to verify and extend his theory. Hertz discovered electromagnetic waves only after he had predicted their existence as a deduction from Maxwell's wave equation. And J. Thomson predicted that the application of pressure would lower the melting-point of ice as a deduction from Carnot's theory of heat.

2. *Control*

(a) *Control over phenomena.* Control in scientific method means the ability to bring about changes in phenomena. The investigator deduces from a law what materials or energies he can expect to manipulate, and the success of those manipulations constitutes a confirmation of the law. The acceleration of subatomic particles to enormously high speeds by accelerators which smash them into other particles in order to disintegrate them and observe the effects constitutes a kind of control over the particles. The control serves as a verification of the known laws and also sometimes as a disclosure of new ones. Scientific laws consists in the discovery of just those elements in the present which the future is bound to resemble. Prediction and control is the testing of the discovery of those laws.

The kind of control which occurs at the present stage in the scientific method, however, is somewhat different. Controls of some kind or other occur, of course, at every stage, and we saw the earliest and most rudimentary evidence of this in the control of observations (Chapter II, 2). The controls referred to in the present chapter are those which are exercised at the most advanced stage where scientific laws exist and are used as instruments for the management of events. Thus advanced control faces two ways: it faces back toward the scientific method, of which it constitutes the last stage; and it faces forward toward practical usefulness, in which field of endeavor pure science is regarded only as a source of technical knowledge. The discussion of technology belongs elsewhere; here we are chiefly concerned with the role played by controls in the scientific method of discovery.

But this much must be said. Technology is engaged in turning the information furnished by the pure sciences over to the service of the practical needs of man and society. So many applications of discoveries in pure science have been found that it is safe to guess that probably there are no inherently inapplicable theories of science which have been checked out by competent professionals. If a law has been established in a science, then there will be a practical use for it. Pure science makes applied science possible, but usually pure science has to be pursued in isolation from all practical considerations if it is to turn up anything which can be applied.

The success of technology has been so great that now we speak of the scientific-industrial cultures. All of the industrial enterprises, from chemical plants to those engaged in the production of nuclear energy, are the by products of the laws which the scientific method of investigation has discovered. Workers in electricity from Galvani and Volta to Ampere and Faraday were concerned with the properties of electric conduction and had no idea that their experiments would lead to the vast dissemination of the supply and control of electricity that is so widespread today. When Liebig in 1840 fed the soil on a plot near Giessen with minerals and so demonstrated that inorganic chemicals could function as fertilizers, he began the enterprise of scientific agriculture from which there have been untold benefits. The production of the fertilizers themselves, of the nitrates for example, was the result of the application of Gibbs' phase rule. The employment of the scientific knowledge gained through pure science in the uses of practical life not only do much to ease the human burdens, they also (and this is more important for the purpose of this book) furnish the last piece of evidence that the laws applied in this way are true.

It must be remembered, then, that control in the sense of practical usefulness does have some relevance to science in the pure sense of basic research. Practicality does satisfy one truth requirement, and that is the test calling on every law for at least one model. In the case of empirical formulas the control of phenomena furnished the minimum model. The application to practical problems of the theorems which can be derived from a law constitutes one more stage in its proof. Here we are looking at practicality from the point of view of support for the law. This is the viewpoint of science proper. Applied science would of course take the reverse view: it would be concerned with science only to the extent to which it could be applied, and for that purpose only. Laws exist for applied science only in so far as they are capable of producing useful consequences. But here we are concerned with verification or falsification for the law itself, considered as a final formulation of the original hypothesis, and this will complete the last stage of the scientific method.

The last test of the law, then, is that which consists in its practical applications. The usefulness of a scientific law, its availability as a tool in the manipulation of relevant phenomena, while not speculatively on the same level of analysis as the three previous tests, does offer it some measure of support. To assume that the practical results which follow from a law absolutely confirm it is to commit the logical fallacy of affirming the consequent. Scientific laws are not true because they work, but work because they are true; and although successful application in practice is therefore not conclusive evidence of truth, still there is no point in going to the other extreme, either, since workability is certainly no argument for falsity. The fact is that workability does furnish some evidence that the credibility as to its truth may not altogether be ill-founded. The theories of nuclear physics were not "proved" by the successful construction of atomic and hydrogen bombs and by the peacetime installation of industrial plants for the production of power, but they were lent a certain strong measure of support in this way.

Control in practice is not, however, precisely what is here meant by control. Control in scientific method as a stage of confirmation differs from control in connection with practical applications if only in the degree of emphasis. The theoretical aspects of control have to do with the scientific tests of laws; the practical aspects of control have to do with applying the consequences of those laws to matters of practical usefulness. In the case of theoretical control the investigator is much concerned with *how* it works, whereas in the case of practical control he is concerned to see only *that* it works. If an antibiotic relieves the symptoms of an infection the practitioner is content, but the scientist will want to know how such relief was effected. Control as practical usefulness may therefore be looked at in two ways: as the degree to which it confirms a law theoretically, or as an efficient way of getting useful things done concretely.

The controls which are put at human disposal to exercise at the mesocosmic level at which individual man moves and has his being come to him chiefly from the microcosmic level. No doubt in the future if the sciences continue to make progress at anything like their present rate discoveries at the level of the macrocosm will also contribute to the control over nature, but in the meanwhile controls from the microcosm are powerfully effective. In physics the knowledge of matter is learned from observations and experiments in quantum mechanics, in biology research in genetics is promising to add to its already strong contribution. The laws are often one integrative level lower than their effects because they analyze or describe the mechanisms at work at the lower levels.

The scientific laws which are employed in the control over phenomena are usually framed in the mathematical language. Consider for example the peculiar relevance of Boolean algebras to electrical networks, and the fact that bridges are built not only by means of the knowledge of technology acquired from previous bridge-building but also by using the calculus of Newton and the analytical geometry of Descartes. It should be remembered in this connection that theorems are applied but axioms are not. The purpose of the axioms is to furnish a basis for the derivation of the theorems, and it is the theorems which are then applied. A mathematically-framed empirical law does not have to be perfect to be useful in application. Chwistek (1948) pointed out that Euclidean geometry is full of defects; its reasonings are not always conclusive, its conventions not entirely justifiable, its appeal to experience questionable. Yet it has been used successfully in engineering for a few thousand years.

All the examples of the control of phenomena show clearly how the theory had to be developed first. The control of material bodies made possible by mechanics depends upon the prior development of the science of mechanics as the theory of the behavior of the material bodies under the action of forces where motion is (i.e. in dynamics) and is not (i.e. in statics) produced. Pharmacology is the study of the control of the human organisms by means of chemicals. We have in the past seen the limitations of an applied science which has not properly developed its theory, in meteorology, for instance, which studies the weather but was weak in predictions and has only lately begun to undertake a measure of control. Now that the physics of outer space has begun to make some advances as a theoretical science, we might expect the meteorological applications of it to be made.

But the successes of applied science hardly need illustration here, they have been so many and so brilliant. The entire chemical industry, for instance, was made possible by highly theoretical work in chemistry, and atomic energy in both its military and civilian uses resulted from speculations and discoveries and experiments in physics of a very abstract nature indeed. There is hardly a department of human life which has not been affected in some way by the practical applications of theoretical work in some one of the experimental sciences.

Prediction and control, like the testing of hypotheses for confirmation in the early stages of the scientific method, are ways of sampling the relevant data. What we have available in the early stages are the data which exist within reach in the present, the data here-now. But the data of today are incomplete; they must be supplemented by those in the past and in the future. History is the account of lost data, but prediction and control have the advantage that they can anticipate the data of the future and deal with them before they pass into the present. Both the data of the past and those of the future are unreliable because unavailable to experiment, as we have noted in an earlier chapter (II, 3(b)). The data of the future can be employed as the basis of a prediction, for unlike the data of the past, they will become the data of the present in some present yet to come. Prediction and control are thus justified in relying upon this final supplement to the incomplete data.

So much as regards time; but then there is also space. All laws have inherent cosmological relevance. They are framed in such a way, as a result of the generalized nature of the mathematical language, that they are presumed to hold everywhere as well as at all times. Unfortunately, the optimism which the investigator is justified in feeling with respect to the future does not apply to the outer reaches of the meta-galaxy; there is reason to suppose that the past will one day be the present, but how is he to sample the data so far removed from his location? Instruments can accomplish something in this direction, but compared to the size of the problem pitifully little.

(b) *The model as control*. Practicality does satisfy one truth requirement: the one calling for a model. A model is a construction to illustrate a scientific law. Models are either mathematical or mechanical, that is to say, composed of abstract objects or of concrete materials. We have discussed the mathematical model already (in Chapter VI, 4), here we shall need to say a few words about the mechanical model, although of course we shall have to include those mechanical models which illustrate mathematical equations. Accordingly here we shall not be concerned with the functioning of models in applied science and technology. Our study of the scientific method however also has a place for mechanical models. The mechanical model is the final test to which the law is subjected.

In mechanical models, materials are employed in little to test the laws according to which they are supposed to operate. A model shows how the laws apply without explicitly referring to them. There exist mechanical models of mathematical equations, as for instance the uniformly suspended cable which is a model of the equation for the catenary plane curve. A planetarium is a model of the solar system, as is any working material structure which follows an abstract theory. The structure of the control over phenomena is a model of the interpretation of the law which is being applied in this way.

Poincaré tried to show that every general theory of phenomenon could be reduced to some form of model. In the nineteenth century mechanical models were in favor, but later these were replaced by mathematical models. The corpuscular model of light was immensely useful even though it has suffered considerable revision. Graphic geometric models are not so much in favor now as models stated in the form of equations, but the older type of model is still in use. Consider for example the Crick-Watson model of the DNA molecule. In empirical science a minimum model is furnished by any practical application.

Models show how laws behave in the material world. A model may be a diagram or an actual three-dimensional construction. Models have always been popular in thermodynamics; the heat engine, in fact, is the very archetype of a model, and has been responsible for many developments. The construction of the atomic bomb or of atomic reactors to produce useful power for peacetime purposes are instances of models. Anything, in short, which is built in the material world after a pattern abstractly described or otherwise symbolically constructed (such as for instance the architect's blueprints), is a model. And the model in this sense is a graphic demonstration that the theory from which it was copied can exercise control to this extent over the material world.

3. *The end of scientific investigation*

We have reached the end of our examination of the scientific method. In the account given in the early chapters, observation first and then induction led to the discovery of an hypothesis. Gradually, supported by an accumulation of evidence, both factual and theoretical, experimental and mathematical, we saw the hypothesis become a theory and the theory a law. Now, supported by the weight of verification, reinforced by its theoretical use in prediction and its practical use in control, the law has reached such a pitch of indubitability as to seem like an incontrovertible fact. Prediction and control are the third and fourth stages respectively in the testing of hypotheses, and the last stage in which the self-corrective nature of the scientific method is able to assert itself. They are repetitive in the sense that they constitute a return to the testing of a theory which was first proposed for that purpose, and at each stage it is still possible to correct an error made earlier, to reverse a finding or to revise a support which has proved erroneous. Thus the last stages in the testing of the results of observations and inductions leading to the adoption of an hypothesis can be construed as evidence that the earlier stages had been worked successfully: the last stages complete and fulfill the early stages.

LAWS IN SCIENCE

Thanks to prediction and control, the law is now as well established and secure as possible. It cannot be denied even by those who bring against it, say, only a single piece of evidence. For in this case the scientist is inclined to reinterpret the falsifying evidence (so long as it is less than the earlier established body of verifying evidence) in such a way that its negative effect is discounted. Not that the law cannot be overthrown – it can; but only by evidence of an overwhelming character, hanging from more than a single chain of inference. The best instance in point is the small positive effects indicating an ether drift obtained with the interferometer first by W. M. Hicks (1922) and later by D. C. Miller and his associates (1933), repeating the Michelson-Morley experiments.

Now that our hypothesis has passed the four tests and changed names twice, what is its status? It has been verified so far as the scientific method is capable of accomplishing this aim. It has survived all the efforts to prove it false, and in the relevant area is now the most reasonable explanation to accept, though not, of course, one established as true beyond recall. Meyerson saw that in explaining phenomena by means of a law, we are to some extent committing the fallacy of synecdoche: accounting for the whole by the part. For each – the law and the phenomena – contains more than the other. Laws extend beyond phenomena as universals over individuals, and phenomena are indefinitely analyzable, containing more than is accounted for by the laws. Moreover, laws are ideal formulations, that is to say, they are stated absolutely, and have to be modified by extenuating circumstances whenever applied. Thus they can always be applied – but never apply absolutely. So far as logic is concerned, it is a matter of choice whether a scientific law is stated universally and approximated statistically, or stated statistically, with universal properties represented by parameters retained as constants in the equations.

We have been dealing with the concrete objects of the material world and with the abstract objects – the laws – derived from them. The scientific tests have involved the relations between them, that is to say, the empirical and the mathematical. If the foregoing account reads as though there were two and only two levels encountered in scientific inquiry: the empirical and the mathematical, and these two quite distinct, it should be borne in mind that these are the outer limits of a situation which is marked by no such clarity at the center. We shall learn as we deal more and more with the scientific findings that the empirical world of reactive forces and the logical world of mathematical relations are quite as distinct as we found them, but that in the endeavor to frame empirically verifiable invariants in the mathematical language other and less well discriminated forms appear.

How do we know when we are working our way toward the discovery of a mathematical law? Tautologies at the empirical level are rare – some would argue that they are not to be found at all. Statistical probabilities are commoner. Yet the aim of the latter is the condition of the former; we only look out for smoke in the hope of finding fire. Perhaps we do not find laws but only hypotheses at the stage of theories, which are still candidates for laws, or probabilities, which are approaches to theories. But we do at least have to suppose that there are laws to be discovered. The more precise and the more abstract our knowledge becomes, the greater the invariantive properties. The empirical sciences are on the right track, at least; and the mathematical sciences offer excellent evidence, physics, for instance.

We have been moving very fast in the exposition, and so we must pause here to take account of certain characteristics and significant aspects of the development by which an empirical proposition becomes a mathematical formula. The key to the transition is the operation of measurement, by means of which we replace less precise qualitative descriptions with more precise quantitative formulations. But the replacement of an hypothesis by a conceptual scheme relies upon the resorting to a set of empirical data represented by propositions or equations. The empirical scientist at this stage of the proceedings literally rearranges his findings to determine whether they cannot be recast into the shape of a mathematical system. This means quite simply whether some cannot be constituted as axioms and other deduced from them as theorems. The more quantitative the empirical formulations, the more easily the mathematical system can be developed.

It should be obvious that such a process cannot be achieved without a good deal of backing and filling. That is to say, the interweaving of inductive and deductive processes must be employed. For the aim of empirical propositions is to attain to the condition of a deductive scheme, but the determination of which among the empirical propositions will best serve as axioms requires an inductive step. We cannot transfer from one language to another without intuitions of fitness. A system is always formulated in a given language which then becomes the domain of the system. In the case of a mathematical formulation within an empirical science, the transfer is never altogether complete, since the symbols will retain some content, i.e., some empirical reference. Our examples are chiefly taken from physics in this regard but this is only because of the advanced development of physics as a science and is not due to any inherently peculiar suitability of physics for mathematical treatment. All empirical material, insofar as its regularity can be properly abstracted, is equally amenable.

Laws are descriptions of the invariability of relationships which enable scientists to predict and control phenomena. Laws, then, as we have noted, are constructed upon the basis of data, and when used for prediction are referred back to the actual world; the movement is from facts through generalizations and back to facts again. Here the principle which Mill introduced of the uniformity of nature is called most vigorously into play. Indeed prediction relies upon the very meaning of the uniformity of nature; for what happened once must under similar circumstances happen again if predictions are to be fulfilled. The prediction of effects is a test of laws which could never be successful were it not for the safety of reliance upon the similarity of the future to the past and the present, at least in some respects and specifically in those respects in which the investigators have formulated a law drawn from the data accumulated in the past but not yet fulfilled in the future.

By scientific laws we mean simply the formulations of the regularities of the behavior of phenomena, abstractly expressed in the mathematical language. It does not matter whether we state these as relative frequencies or as immutable laws so long as they have a sufficient degree of empirical support. To be ideal is to be without exception, yet we can scarcely hope that any law will attain it, and in any case this could never be made certain. There are always propositions which it is claimed transcend experience – religious revelations, for instance – as opposed to those propositions – such as scientific laws – which are held accountable to experience. The opposition is a false one; for while it is true that revelations cannot be checked, yet what scientific method does is to tell us of the validity of allegedly universal laws. For universal laws will be laws perpetually operating. Thus what transcends experience – a scientific law, for instance – may yet be held accountable to experience.

The presumption of generality on the part of a law of science always goes well beyond anything that may have been proved. All that is meant by "proof" in this connection, therefore, is the discovery of such evidence as would make tentatively acceptable an hypothesis (later a theory and finally a law) which had been presented for testing. Submission to the twin ordeals of prediction and control are only further stages on the way to establishment; they furnish a certain measure of support and offer firm evidence of invariance, but they do not certify a law as absolute, even so. They do mean, however, that the law may now be used for both theoretical and practical purposes pending the development of evidence to the contrary.

Mention should be made here also of laws which exist in the sciences without sufficient proof but which are useful and widely applicable, and to which no exception has yet been found. The exclusion principle of Pauli (1925), according to which no electron in an atom can simultaneously have the same spin and occupy the same orbit, has been immensely useful in chemistry where complex molecules containing large numbers of electrons still come under the principle. Another example of such a law is Wien's law, that the hotter a surface the bluer the light by which it shines. The support for such laws comes mainly from their pragmatic value but there is some support also from their consistency with the existing body of knowledge in the same science. Many such partially unconfirmed laws exist throughout the basic sciences and are in continual use.

Laws, then, remain on semi-probation. As long as a law exists, there could be future occasions when events of the kind described in weak deductions from it could be expected to be fulfilled. Thus it can be regarded continually as established only *pro tem* and under conditions which render it susceptible of being brought again and again to trial. Laws, no matter how firmly established, are provisional always, and are entirely subject to revision and even abandonment when challenged by some fresh discovery. The discovery in question may be that of a difficulty. It was the shortcomings in Newtonian mechanics which did not allow for certain phenomena, such as the results of the Michelson-Morley experiment and the Fitzgerald contraction, which led to Einstein's special theory of relativity. Difficulties encountered in applying the laws of classical physics to the model of the atom devised by Rutherford, according to which electrons should eventually be drawn into the nucleus, led Bohr to postulate energy levels within the atom and to assume that Planck's quantum was involved in any change in levels, the electron jump being accompanied by an absorption or emission of energy in fixed quanta.

Laws are never absolute. In scientific method no conclusions are ever reached which are then placed beyond the legitimate range of further inquiry. Any law may be called into question at any time. Only the axioms and the method of science are constant. The laws discovered by means of them are subject to revision or even abandonment should the evidence ever require it. Laws are merely the results of persistent investigation. The survival of a scientific law through various testings by means of the prediction of effects and the control of phenomena is merely a signal that for the time being further inquiry in that direction is not called for.

It may be called for again at any time that new evidence warrants. For scientific development, like most other activities and events, proceeds according to a rhythm: discovery, establishment, difficulties, re-examination, re-discovery, re-establishment, in an endless cycle. Continuity of progress is seldom a straight-line development; there are periods of feverish activity followed by periods of comparative calm. Advances in knowledge are made in spurts, first in one science and then in another, and the tide in each one recedes in periodic beats. A law which survives one period of positive activity in a science may well be the victim of the next, meanwhile having been appealed to in the interim by theorists and practitioners alike as one of the cornerstones of the entire edifice. Thus even prediction and control carry no sanctification; but they are the final endorsements made available by a procedure in which no generalizations are ever altogether safe.

Looking back over the centuries during which science has been in operation, we can see a gradual, cumulative gain. It does not seem possible to chart the future course of a science as a whole. The tradition goes in whatever direction successful investigation may take it, and a continuity may be read only by the facts and not by any forward extrapolation. The sum of scientific knowledge does not remain a constant but increases by jumps – whenever in fact there is a breakthrough – so the residue of findings accumulates and furnishes a kind of mutual reinforcement to build an explanation and accounting of the world which at least has completion as its aim even though never attaining to that condition altogether.

At the last stage of prediction and control it is possible to look back and see that there has taken place an increasing degree of abstraction from stage to stage. Even though it is possible to recognize in a law the theory which led to it and the hypothesis which led to both, the theory is often more abstract than the hypothesis and the law more abstract than the theory. The stages of science when put together constitute a structure stronger than any one or all of the stages considered separately. For now there is clearly in evidence the logical structure of a single method. The aim of science is to push logic as far as it can go and yet remain material, to generalize but always about substance. The stages may be pursued separately by different investigators; it often happens also that a single stage is sufficient to account for a particular scientific development. Yet the fact remains that even in such cases there is the exception operating against the background of a method. Those who have argued for a number of scientific methods: as many as there are distinguishable sciences or, even worse, as many as there are scientific investigators, have reckoned without the relations between the stages and have argued only from the extenuating circumstances which occur in each case or from the differences appropriate to each separate science. There is only one scientific method and it is stated for logic, and there are always extenuating circumstances, just as in the case of Boyle's law for ideal gasses, which has to be modified when the specific temperature and pressure is known in any particular instance. In this way it has been possible to show that there is a logical structure to scientific investigation, and not a mere random collection of ways of testing theories by means of facts.

We have been looking at the scientific method in some detail, stage by stage. This task is now completed. The entire aim of the scientific method is of course to produce findings. These are of more than one kind. In the next chapter we shall be examining the variety of findings in detail.

CHAPTER VIII

TYPES OF EMPIRICAL DISCOVERIES

When the word "science" is used, it is not always clear what is intended, and a confusion often results from failing to distinguish between a *technique* of inquiry (the method) an *area* in which inquiries are conducted (the domain), the *results* of such inquiries (the findings), and the *equipment* of the professionals whose business it is to pursue such inquiries (the enterprise). Science as such is characterized by its method, and so by discussing the method of science we should be engaging in the understanding of science. But here instead we shall undertake to discuss the findings of science: the laws and other objects found in the domain by the sciences, which have received far less attention.

Despite the disproportionately small amount of space given over to the discoveries of science in works devoted to the discussion of the scientific method, it is for the sake of making discoveries that the scientific method is practiced. It has often been said that the greatest discovery in science was the discovery of the scientific method of discovery. This is no doubt true. Only, here we are regarding that method as a topic which has been investigated, and we shall be more concerned with the consideration of the type of discoveries which are made with its aid.

But there is another and much more neglected discovery in science which should rank equally with the discovery of the method itself. It is seldom mentioned that every great discovery in science, whether it be of an entity, a process, or a law, is accompanied by the recognition of some new area of ignorance. For instance, we now know that there are short range forces which hold together the particles of the nucleus of the atom, protons and neutrons particularly, and that these are divided into weak and strong interactions, but we do not know of what such forces consist. In a word, what has been discovered is a new area of ignorance. New areas of ignorance have been discovered with the discovery of subconscious mental processes in abnormal psychology and the discovery of the galactic nucleus in astrophysics. New areas of ignorance are enormously helpful, of course, because they give rise to new types of investigation. But the fact remains that the claim to progress in the understanding of nature has to be modified by the degree to which the extent of the ignorance exceeds the knowledge increased by the new discovery.

To learn where a science is in periods of great advance, it is necessary to keep up with the journals. The findings change from time to time and are replaced by other findings, but in the present state of the physical sciences these changes are recorded rapidly. There always remains over from any series of tests in which an hypothesis is promoted to the status first of a theory and then of a law, a component of uncertainty which holds it forever as a proposition, with something of the nature of the hypothesis still clinging to it. It is never altogether free from suspicion, never safe from the threats of modification or abandonment, never altogether established. At any time, its explanatory value may prove to be misleading or too narrow or in some other way inadequate; and then it will have to be replaced.

What we wish to examine here, then, are not the discoveries of science, for this would be the task of a scientist, but the various *types* of discoveries. What kind of result is likely to be found among the end products of scientific investigation?

The demarcations between the findings are not absolute; a science will be occupied with the relations between systems and laws as much as with the systems and laws themselves, with the transitional types between entities and processes as well as with those entities and processes. The subject-matter of science is always presented as though it were deceptively simple, whereas it is indeterminately complex. As much could be said for the unwarrantedly flat account of the scientific method whereby empirical discoveries are made. The description of the method usually given is of a linear structure of a sequence of activities, a logical procedure laid down along a time line. But nature is everywhere dense, with a qualitative richness which does not allow such simplicity to prevail.

Thus at every stage in the career of an investigation there are side issues, tangential considerations, incidental effects, unsuspected areas. It should never be forgotten that the scientific method is an *activity*, in which many of the sharp and fast distinctions which are so familiar actually serve to conceal the rich set of interdependencies which are unfolded in the course of an ongoing process. Consider the tight system of relationships in science, for instance new theories making possible the construction of new instruments, and new instruments opening up new empirical areas, in which in turn new laws are uncovered.

We shall consider briefly eight types of discoveries. These are empirical systems, empirical areas, empirical, causal and statistical laws, entities, processes, formulas and rules, and procedural principles.

1. *Empirical systems*

We shall be concerned with systems only so far as they result from empirical investigations. In an empirical system, laws are arranged in such a fashion that a number of them (the theorems) can be derived from the assumption of a few (the axioms). Laws in general are descriptions of the data of experience. They are systematic rather than summatory. The procedure of enumeration simply will not do for description in the sciences, and a shorthand method is needed. Laws are one answer; the axiomatic method is another. The axiomatic method furnished those large assumptions against which the laws are meaningful. Empirical systems in this sense often are called theories. (The terminology is not standarized.)

By an empirical system here, then, is meant an abstract system interpreted by substituting for the variables symbols representing concrete entities or processes. An empirical system is a deductive structure of experimentally verified equations which are consistent, a few equations serving as the axiom-set for the derivation of the others – Newtonian mechanics, for instance, relativity mechanics or quantum mechanics. It summarizes results in a convenient fashion which enables investigators to test them more readily, an orderly though provisional arrangement for purposes of examination. For an empirical system defines the conditions of a hypothetical domain which it is expected will accord more closely than any other to the world as disclosed by ordinary experience.

Not all scientific findings are empirical systems, it should be remembered. Such systems represent the more comprehensive findings in the sciences, and usually they are the result of carrying through to the very last of all of the seven stages in the scientific method. Empirical findings may be spun off much earlier, however, in fact at any stage in the process. Not every inquiry goes through all the stages; not all start with observation, and many end before prediction and control are reached. Mathematical rigor is something of a rarity in empirical science despite its widespread use of mathematics. When propositions which have been empirically derived are arranged in such a way that some serve as the assumptions of an axiom system, the system may be as rigorous as a mathematical system. Most empirical systems constructed on the axiomatic method are, however, partially formalized systems.

The chief value of an empirical system lies in its power of generalization.

In effect it sums up a vast amount of observing, experimenting and theorizing, and produces to a greater degree than would be otherwise possible predictions of events and control over phenomena. It may open up empirical areas to investigation, it contains laws as elements, and it allows for the deducibility of entities and processes, formulas and rules. It functions as a large organizing force, and it clarifies the operations in a given area of investigation. Its systematic character is comprehensive, and for a while at least it appears to be complete. Eventually this is found not to be the case. If observations and experiments are continued, the limitations of any empirical system usually begin to appear.

No matter how inclusive and rigorous an empirical system, the informal, the rich and the partly irregular character of the world disclosed to experience will always break through. Thus empirical systems are systems in so far as they are mathematical, and unsystematic in so far as they are empirical; and so an empirical system represents a compromise finding, an empirical situation interpreted by means of a mathematical system. That empirical systems are genuine findings is shown by the opposed facts that on the one hand they are predictive and on the other hand such predictiveness does not imply an absolute determinism. Given the initial state of an isolated set of physical conditions and its rate of change as expressed by differential equations, then its state at any future instant can be determined. But that such predictiveness is partly indeterminate follows from the ideal character of complete isolation and the limits of isolated systems. No actual systems are ever altogether isolated or absolutely inclusive.

It may therefore well be asked to what extent an empirical system can fairly be characterized as a finding in a science, even though it is clear that this is the result at which the science has arrived. It is necessary that the reference of universals extends beyond experience, and so no scientific proposition has ever been wholly in accord with experience because scientific propositions are perforce universally stated. "This patient suffers from cardiataxia" is a statement about an individual, but it employs no less than five universals. It places a given individual under a term whose intention defines it as a class and whose extension is unknown. The universality contained in all scientific terms, whether constants or variables, is inherent in the nature of representation and consequently a property of all communication systems. There are names for some individuals, but science is rarely concerned with them. Science is concerned primarily with those classes whose members have been and can be disclosed to experience, and with the relations between the classes, that is to say, propositions. If next by distinguishing in a set of propositions those which can serve as axioms from others which are deducible as theorems the propositions are combined in such a way as to construct a system, it is clear that the investigator has multiplied the assumptions, which now include not only the universal nature of the propositions but also the unverifiable nature of the empirical system itself. But it is the phenomena of the conformity of a segment of existence to an interpretation of the empirical system that allows us to regard that system as a discovery.

2. *Empirical areas*

The scientific method is a way of probing the depth and extent of the empirical world, dimensions which are by no means always readily evident to ordinary experience. Surprisingly enough, there are actual empirical areas, segments of existence serving as arenas for the inter-actions of material entities and processes in space and time, which remain unknown until the scientific method at some stage in its investigation uncovers them. It often happens that new conceptual schemes can be used for probing the areas, as can also – and more obviously – new material instruments. Let us look at some examples.

Charcot's conceptual scheme together with the evidence of a single patient suffering from conversion hysteria suggested to Freud the possibility of opening up a whole new empirical area to investigation: that of the unconscious mind. Charcot's studies in the relationships between hypnotism and hysteria influenced both Janet and Freud; but whereas Janet went on to consider the concept of the integration of the personality, Freud, influenced also by Breuer, was concerned with the interpretation of disintegration as the disturbance of equilibrium, and its restoration therapeutically by induced memories to eliminate conflict. When Freud substituted free association from hypnosis and recognized the role played by sex in many disturbances, the method and its area of investigation were so to speak conceptually established. The results obtained later by hordes of investigators have confirmed the existence of an unconscious mind and have begun the exploration of its contents. That bio-chemical methods have been added to psychological methods only serves to reinforce the evidence for the existence of such an empirical area.

The development during the second half of the seventeenth century of a microscope with a single lens of considerable power provided an instrument which brought into focus the area studied in microbiology. In the hands of such investigators as Boyle, Swammerdam, Malpighi, and Leeuwenhoek, the microscope made visible the world of tissues and organs. Microorganisms in general, bacteria, spermatozoa, as well as red blood corpuscles and cells, became known because observed for the first time. Thus the microscope played a crucial role in the discovery of an empirical area. Before the invention and use of the microscope, such a world had not been unknown to exist; since then the scientific knowledge of it has been paced by the further development of the microscope, the most recent example of which is the electron microscope with its magnifications counted in hundreds of thousands.

The combination of conceptual scheme and instrument is dramatically illustrated by the discovery of the quantum theory. Rayleigh's law of radiation called for a monotonic increase in spectral density with frequency in a state of equilibrium, in strict accordance with classical mechanics. But experiments showed that in black body radiation the spectral density decreased after a point even though the frequency continued to increase. Planck was concerned to explain the discrepancy, and he did so by assuming a discontinuity of radiation: that matter can emit radiant energy only in finite quantities which are proportional to the frequency. A formula was obtained which coincides with Rayleigh's law for low frequencies and high temperatures but not for high frequencies and low temperatures. The finite quantities are multiples of a constant of action, Planck's quantum, h. An atom can change its energy state only by receiving or emitting quanta. A whole world of phenomena, not inconsistent with thermodynamics but definitely inconsistent with classical mechanics, was opened up, the world of quantum phenomena, studied now by that division of physics called quantum mechanics.

In addition to the probings of conceptual schemes and mechanical instruments, or, what is more common, combinations of the same, there are other ways in which new empirical areas are opened up to investigation. One way is by the splitting off from or the dividing of older empirical areas.

A good example is to be found in the physical science of crystallography. Already in the sixteenth century Agricola had used the facilities of the mining industry to study the properties of crystals. In the second quarter of the seventeenth century, Boyle studied crystals in search of the structure of chemicals. But crystals were still associated with geology. For many years thereafter crystals were of scientific interest only for their light refracting properties and so aids to the advance of optics. In the hands of Herschel, Berzelius and Pasteur in the nineteenth century, crystals played a role also in organic chemistry. In 1880 Curie discovered the piezoelectric effect, that crystals if squeezed will emit an electric current while changing shape. At the turn of the last century von Laue and the Braggs had studied the properties of X-ray by using crystals as gratings. When, then, did the science of crystallography achieve some measure of independence? With the discovery that crystals could be understood better by means of the mathematical theory of lattices? Or with the discovery that crystallinity is a property of the solid state of matter? In any case, it is clear that crystallography as a separate science is the child of other physical sciences: chemistry, optics, and radiology, with the assistance of geometry.

Additional examples are not far to seek. Recent work on the reticular formation of the brain stem by Magoun and others has laid bare an important area of investigation in the central nervous system. Again, the exploration in the last century of the effects of stimuli on consciousness, have opened up an entirely new area of inquiry in psychology, that of psychophysical correspondence.

The result of uncovering empirical areas is of course to stimulate discovery. This is usually accomplished by means of new problems. For instance, the nature of the powerful short-range forces which have been found in the nucleus of the atom remains thus far unknown and so constitutes a problem. Before the development of atomic physics, this problem as a problem did not exist. One kind of progress in science, then, consists in the discovery of important problems. But this is not the only kind. Many new laws, hitherto unrecognized entities and processes, formulas and rules have been unearthed in freshly discovered empirical areas. Scientists are forever observing, making inductions, adopting hypotheses, deducing from their hypotheses to relevant empirical areas, experimenting, making inductions from empirical areas in support of their hypotheses, calculating mathematically from theories, predicting from laws and applying controls. The dialectical process turns over once more, and the empirical feeds the theoretical as much as the theoretical in turn feeds the empirical. The application of the scientific method in a particular science is thus given a powerful stimulation.

3. *Laws*

Laws are the correct way of expressing specific uniformities in the behavior of natural phenomena. Finding a law in an empirical area is like inserting a knife where there is a natural break. It recognizes as well as establishes cleavages. Laws in science vary with respect to their degree of abstractness. There are laws expressible as functions: water freezes at 0°C., for instance. These are the results of repeatable experiences. Then there are the general laws, which are constituted by observed regularities: all men are mortal. Next there are the abstract laws: the law of gravitation, for instance.

Laws are universally stated. For instance, that heat cannot be transferred from a body at a given temperature to a body at a higher temperature (second law of thermodynamics), as for instance, when we attempt to transform heat into work and find that it is not the same as when we transform work into heat. But laws are also exemplified only in individual instances, and it is well known that no finite number of individuals exhausts a universal. How is the dilemma to be resolved? It comes to something like this. To say that we recognize a scientific law is to say that we know how the relevant phenomena within certain limits will behave anywhere and at any time given the extenuating circumstances. A law, then, is nothing more than a description of the behavior of material objects or of the average of what situational behaviors will have in common. But this does not deprive the law of its logical status as the formal expression of empirical information, for we can look at it also from another point of view as a logical invariant filled with empirical content.

There are many varieties of laws in science; we shall deal with the three main ones: empirical laws, causal laws and statistical laws.

(a) *Empirical laws.* At the lowest level of generalizations, the laws based on observed regularities which do not extend beyond the available evidence in terms of a theory, are called empirical laws. Empirical "laws" are not laws at all but summaries of the tendencies of phenomena. An empirical law is a symbolic representation of a range of phenomena. The meaning of the law is the relationship exemplified in the phenomena, and the evidence for the law must have been found in the phenomena, or, and this is the devastating part, in such samples of the phenomena as have been brought into evidence.

Empirical laws have their place in the scientific method of investigation, but they are not normally considered ideal in science because they are not sufficiently abstract. In so far as they are confined to the phenomena they do not have the generality of law, and in so far as they have the generality of law they are not confined to the phenomena. And when they are not confined to the phenomena they are not empirical. Empirical laws are, as Herschel said, univerified inductions, to be treated with the utmost reserve. For science requires, so to speak, that they be confirmed by theory.

An empirical law is a law having some content derived from a scientific domain, from a physical, a chemical or a biological domain, for instance. That they require a theory is compelled by the inherent nature of data. Not only are all facts related but it is assumed in science that they can be explained. Thus an empirical law is a starting-point for investigation, not a finished scientific product.

There are degrees of empiricism in empirical laws, however. In some the signs of generality are more evident than in others. There are for instance empirical laws which describe events in terms of certain fixed relationships, without ever assigning causes. Starling's law of the heart, that the force of a contraction of the ventricular myocardium is proportional to the degree of ventricular filling by venous pressure preceding the contraction, is an example. In this degree of empirical law, we are told *how*, not *why*, certain events occur.

Empirical laws of whatever degree have in common the failing that they do not probe deeply enough. No collection of formulas in science can ever exhaust the phenomena, so that laws are never more than approximative. Even where they are accurate, they are never complete. Now abstraction, as we shall see, is a way of probing, but empirical laws are never sufficiently abstract. If it is not possible in framing empirical laws to go beyond the statements confined to observation, then the truth so far as it is sought in science by means of a correspondence with facts is forever beyond discovery. Fortunately, there are further degrees of abstraction.

(b) *Statistical laws*. Statistical laws are descriptions of the behavior of phenomena which are not entirely determined, laws derived from the study of large populations of instances. A good example is furnished by the current statistics on the relation between the consumption of cigarettes and the incidence of lung cancer. Other examples are to be found in vital statistics, and in the laws of heredity in genetics. In quantum mechanics, in contrast with the classical mechanics, it is not possible to calculate the position of a particle from its initial conditions and velocity. Repeated observations are necessary in order to give us the approximate position. In other quantum phenomena we encounter statistical situations; for instance in the study of the discrete energy states of radiation oscillators and in the wave-particle duality. Statistical laws are never complete and always allow for exceptions; therefore, although derived from populations consisting of individuals, they are not applicable to individuals. If, for instance, it is found that the mean for the inhabitants of a given neighborhood in some particular city is 2.7 persons per house, it is obvious that despite the accuracy of this figure there is not a single house containing 2.7 persons. Statistical laws are laws descriptive of the properties or of the behavior of populations of individuals. These properties or this behavior may be different from those of the individuals composing the population. For the population is more than the sum of its individual members and exhibits characteristics not possessed by them. The mode or the median of a given population may represent a quantity which does not hold for its individual members, and the standard deviation may measure the dispersion around a wholly *ad hoc* mean.

Statistical laws, which measure the relative frequency of a particular characteristic in a given population, are laws dealing with classes usually finite and at most trans-finite. A trans-finite class is here defined as a class with a large membership which is possibly infinite. They are distinguished in this respect from causal laws which are laws dealing with classes having an infinite membership. Some events are repeated with relative frequency, others invariably; the former are candidates for inclusion under statistical laws, the latter under causal laws. Statistical laws disclose tendencies; causal laws state determinisms. Statistical laws are related to causal laws by the existence of hidden parameters underlying the equations. In short, statistical laws are the laws describing the effects of the multiple causes which are susceptible to small variations. They show the forms in the last vestiges of irregularities left in the facts.

Statistical laws are the raw material for mathematical probability, which drives a wedge between unexceptional propositions on the one hand and tendencies on the other.

(c) *Causal laws.* The final degree of support for an empirical theory, then, is that accorded by the status of causal law. When a theory is framed in terms of causality, then we may suppose that to investigators it has achieved all the verification that mathematics can lend to it. For causality is consistency – the consistency of classes of material objects. Causal laws describe unchanging elements, the evidence of logic at work in the actual world.

A causal occasion is a logical occasion.

If *A*, then *B*

is a statement in pure logic, but the assertion that

whenever *A* happens, then *B* will happen

makes it clear that we are dealing with something else, with cause-and-effect, or with applied logic, in a way in which space, time, matter and energy are required for there to be happenings. Causality is no mere sequential juxtaposition; it is a logical nexus.

Examples of causal law abound. In physics, a favorite is the law of entropy. In chemistry, the case for causal laws is even stronger: every chemical formula constitutes a separate causal law. In general, causality is expressed by the relation, "if . . ., then . . .," and every causal situation is an example of material implication. Causality is the empirical correlate of logical consistency and therefore while always stated as a logical ideal, never ideally carried out. The ideal of causality is mechanism. For a mechanism represents a physical interpretation of the *if-then* situation. If such and such steps are taken, then such and such other steps will inevitably result. It is never the case that an internal combustion engine is built and furnished with fuel and a starting mechanism, and an output of power is not the result.

Hume tried to demonstrate that the belief in causality rests on custom rather than on the evidence of the senses, but Kant saw that it suggested a more intimate relation between the logical and the empirical than Hume had supposed. Only, where Kant thought to reconcile logic with empiricism subjectively, the scientific method does so objectively. Causality and mechanism describe the effect of logical implication in material events in space and time. Mill described causation as an invariability of succession which obtains between every fact in nature and some preceding fact, but he failed to weigh seriously the implications of invariability. Such an absoluteness of relationship involves a logical connection between natural events, which is all that the proponents of a real causality require.

The attack on the validity of causality as a property of the world has come from two developments. It has come from the evidence of Heisenberg's indeterminacy principle, and it has come from the theory of probability. The first attack is a rather complicated one, and we shall say just a word about it here. A reply to the second attack will be made shortly.

Let us begin by starting briefly the terms of the first attack. Heisenberg's principle of uncertainty states in effect that we cannot measure both the position and the momentum of an elementary particle. The specification of either must make the specification of the other impossible because of the implied ascription of contradictory properties. Heisenberg's principle was enunciated as a result of the limitations of observation at sub-atomic levels: the radiation used to determine the location of a particle will interfere with it; as the frequency is increased, the momentum is affected. Wave electrons tell us nothing of position; corpuscular electrons tell us about frequencies. In other words, the observer carries his limitations around with him; and they are so severe, we are told, that man becomes, in Bohr's words, an actor as well as a spectator in the drama.

The contention is full of difficulties. In the first place, the datum, which is undeniable, has been interpreted subjectively in a manner which is unwarranted, for a methodological limitation has been erected into a natural constituent. It may be that we are unjustified in attributing to the external world a failure which we find in our own abilities to investigate it. There is an indefensible arrogance in the assumption that the whole universe suffers from human weakness. In the second place, the limitation is one involving the instrument of observation rather than the subject observing. Presumably, another instrument or another line of investigation might get round this difficulty and so dissipate the entire question. The qualitatively infinite richness of nature is not to be characterized and bound by the human inability to go beyond its own inherent shortcomings in the way of investigation. Moreover, as new concepts and new techniques are devised for conducting research, the present set of limitations may well evaporate.

The early advocates of causality encouraged in their enthusiasm by the success of the scientific method, have gone too far. Laplace thought that the range of causality was without limit, that the laws of classical mechanics held without exception and so applied absolutely throughout nature. But Bohm has argued that mechanism as an explanatory principle is strictly a function of isolated systems, which is to say limited systems, and therefore cannot be applied to the universe as a whole. There are reasons for supposing that Bohm is right. For one thing, the existence of the integrative levels, where a mechanism at a given level will produce the expected results only at one other, would be a sufficient illustration of the limitations of mechanism.

For another thing, in science there is a pretty widespread tendency to reduce the area of application of any particular explanatory principles by uncovering errant cases which require more general explanatory principles. What is happening to the theory of mechanism has already happened to Newtonian mechanics. Investigators progressively discover that the facts are more diverse than they had supposed and the general principles more limited. For instance, estimates of the age of the earth and of the size of the universe grow steadily larger. Doubts concerning the cosmic applicability of terrestrial principles grow with them. Causality meets one of the criteria of logical systems but not both; it meets the criterion of consistency but not of completeness. Not every scientific formulation is either demonstrably true or inconsistent with some causality expressed statement. Clearly, for a total explanation, then another principle is required.

In the scientific subject-matter, it is usually the case that cause-and-effect linkages are not found in isolation. Causal sequences in nature are the rule, and linear series of causes and effects can usually be traced. Cournot has indicated the importance of separate causal chains. Every event is included in some causal sequence but the causal sequences are not necessarily related directly.

Causal laws are descriptions of forces at work in nature but are not themselves those forces. The error of supposing that causal laws are laws that cause certain things to happen relies upon the linguistic fallacy which takes the word for the deed, very much as was done in early verbal religious magic, where the name of the god could either compel the god or invoke his wrath. Causal laws are descriptions of behavior which is invariant given the initial conditions. We have learned that energy becomes less and less available not because there is a law of entropy but because the law of entropy is an accurate description of the degradation of energy.

Causal laws, then, are absolute statements suggested by an examination of some segment of nature. They have a veridical ring about them which is both partly true and partly misleading. It is partly true because in the work of investigation they are candidates for the position of tautologies. It is partly misleading because it assumes a certainty of attainment, where the conditions warrant uncertainty or at the most a very high probability. In short, a causal law is one which is framed in terms of absoluteness, but where the evidence for its completeness remain itself incomplete.

4. *Entities*

The scientific findings may consist in the discovery of something as concrete as an empirical area but much smaller in extent: an entity or a process.

Entities are classes of material objects. Examples of such entities are: electron, uranium, trypsin, neuron, tumor, personality, shaman, matriarchy. Investigators do not permanently recognize the existence of such classes unless at least some members are disclosed to sense experience and the experience lends itself to repetition. If the working theorem be accepted that there are no unique material objects in existence, then the experience of one such object is assumed to be sufficient evidence for the class. On the prior principle of economy, that no more entities should be admitted than are necessary, the experience with a single entity may not be taken as evidence of anything more than that material object. But the aim of science is not the discovery of material objects but of classes of material objects and of laws concerning classes and eventually of empirical systems. And so the working theorem that there are no unique material objects in existence, when combined with the discovery of a single material object, might be sufficient evidence for the class of such objects. This was perhaps the situation with respect to the planet earth at the turn of the century.

Experience with a single entity is rare; and usually where there is a material object disclosed to experience, other objects similar in type may be found. In the far more common case that a number of material objects are found which were not known before but which exhibit common characteristics, it is clear that the scientific method is aimed at the discovery of classes for which there is empirical evidence and not with the material objects themselves except as they constitute the evidence for such classes. The direction of empiricism is always toward the lowering of empiricism but only when there is sufficient empirical justification.

Not all the scientific entities are equally precise. Of the ones we have enumerated, "uranium" is more precise than "electron" and both are more precise than "personality." Intermediate types can be found. The "field" in physics, "intrinsic energy" in thermochemistry, "sclerosis" in medicine, and "aptitude" in psychology. An entity so conceived is called a datum, the considerations of its width and depth are now relevant; that is to say, how much does a newly discovered datum cover, and at what level of analysis is it to be found?

The width of a datum is exceedingly difficult to estimate and all but impossible to define. How much does the datum include: what exactly does it cover, and how suggestive is it? Consider for instance a very complex entity, say DNA, the double helix whose role in genetics was discovered by Watson and Crick in 1953. DNA is the primary coding material which is found in some of the viruses and bacteria. It can do what some genes are known to do: replicate, transfer information to other parts of living systems, and undergo mutational alteration. What are the limits of homogeneity of "phase" in chemistry, of "disease" in pathology, or of "neurosis" in abnormal psychology? And in anthropology what are the boundaries of a "society"?

It is just as difficult to determine the depths of an entity, for there we have the two concerns: level of analysis and perspective. An analytical level is an integrative level viewed from the level above. A molecule is a combination or synthesis of atoms at the chemical level, but an analytical element of the cell. The sharpest demarcations are the level demarcations, however they may be viewed. For given the determination of perspective, each is fixed. At what analytical level is the entity observable or from what perspective can its existence be inferred. A proton is an analytical element only from the perspective of ordinary experience; from the standpoint of world conditions, it has the same kind of existence as a tree or a galaxy.

Most of the entities of empirical science are observable, but not all. There are marginal cases, such as the hypothetical entities, which are entities proposed in order to account for observations. The "neutrino" and the "meson" in atomic physics are hypothetical entities, as is the "gene," the unit of inherited material in the chromosome, in biology, or the "ego" and the "id," with the superego, the entities of the unconscious in psychoanalysis. The planet Neptune was at one time a hypothetical entity and so was phlogiston; neither is so any longer. Neptune has been observed and is now a well organized astronomical object, while phlogiston was dropped from the class of acknowledged entities in chemistry after its refutation by Lavoisier. Neptune was a hypothetical entity in the short time between its hypothetization by Adams and Leverrier in 1840 and its observation by Galle in 1846. Phlogiston was an hypothetical entity from its postulation by Stahl and others in the middle of the eighteenth century until its refutation by Lavoisier by means of the balance in a publication of 1777. In that year, however, Scheele was still supporting the phlogiston theory by maintaining that in the photographic process silver chloride was converted into silver by its ability to derive phlogiston from light. Generally speaking, the hypothetical state is not a permanent one, and it is expected that the true condition of an hypothetical entity will be ascertained one way or the other; then either it will pass into the class of observables or be eliminated as not being an entity that scientists take seriously.

One familiar variety of hypothetical entity is the ideal entity. An ideal entity is a logical set of limiting conditions suggested by material objects. Examples are: "black bodies," "frictionless engines," "instantaneous rates," "perfect vacua." One definition should be sufficient; a "black body," for example, is one which absorby all radiation falling upon it. Ideal entities have the status of logical objects; that is to say, they certainly do have some sort of being, just as mathematical objects have, but it is not that of material objects. The exact status of limiting conditions has been the topic of much controversy. It has never been settled satisfactorily; yet its consequences are of vast importance. Scientific investigators employ such entities regularly; they are needed in dealing with actually existing entities. Black body radiation, for instance, is a standard for the comparison of the absorption of radiation of all frequencies by given material bodies.

Perhaps with the introduction of the empirical model the widest datum becomes indistinguishable from an ideal entity. A model is the blueprint of an entity which could conceivably be either constructed or approximated in practice. Just how seriously does the physicist take his models of atomic nuclei, the "powder model" resembling statistical mechanics and the "shell model" resembling a tiny planetary system? Each has yielded certain useful conclusions. One model which has had the longest history in empirical science is Poisson's statistical models of urns containing balls of different colors, from which samples may be drawn.

The decision as to the status of an entity, hypothetical or ideal, may be a very delicate business and one not easy to make. At what point can it be decided, for instance, when an entity is an entity and not a process? Everything in existence is involved in continual change; so that an entity is a process considered as arrested, just as, we shall see presently, a process is an entity considered as active. If a definition has to be made in terms of similarity and difference, then for any entity the similarity is provided by its classification and the difference by its particular activity. Thus the photoelectric effect has been considered a quantum of radiant energy, the wave

intensity of light being a matter of the concentration of photons. From the point of view of change, then, an entity may be described as a structural process.

5. *Processes*

We have just noted that in a certain sense there are no empirical entities which are altogether static. Only logical or mathematical entities justify that description: "circle," say, or $\sqrt{-1}$. But there we are talking about empirical entities, and at most these are short-range processes. An empirical entity is what it does. This can be made clearer perhaps by citing as examples some so-called entities in which the process aspect is uppermost. First of all, there are entities indeterminable with respect to their status as entities or processes: the "quantum constant," for instance, or De Broglie's 'matter-waves." Then there are processes ordinarily conceived as entities: "catalysis" in chemistry, "metabolism" in biology, or "habit" in psychology. Finally, we have the clear-cut processes: "cascade decay" in atomic physics, "elasticity" in mechanics, "eclampsia" in biology, "acculturation" in anthropology.

A process is a class of structured activities, a way of effecting a change of states. Like entities, a class of processes is not permanently recognized as such unless at least some members are disclosed to sense experience and the experience lends itself to repetition. It is possible to repeat instances of the electrolysis of water at will.

The integrative levels so far as process is concerned are analogies of the energy levels within the atom. And any entity can now be described in terms of process as an output of energy; more specifically, as a sensitivity-reactivity system at various energy-levels. So, for instance, for an electron, a molecule, a cell, an organism, a society.

Activities which produce changes are called processes whether or not they are effected by an external agency. Radiactive disintegration is a process which does not depend upon an external agent to bring it about; agriculture which employs dissolved minerals rather than soil for the cultivation of plants (hydroponics) is a process which does depend upon an external agency. Many so-called laws are nothing more than descriptions of process, such as Graham's Law, which states that a gas diffuses with a velocity which is inversely proportional to the square root of its density. The most famous process now is that of atomic energy: the transformation of matter into energy through the loss of a small amount of mass and the release of a large amount of energy.

Finally, there are certain practices which are known as processes, currently employed in applied science and industry, such as the "lead-chamber process" in chemistry for the production of sulphuric acid, the "open-hearth process" of steel production, or the Mond process for extracting nickel by employing the action of carbon monoxide on the ore to yield a gas which when heated decomposes into pure nickel and carbon monoxide.

6. *Formulas and rules*

It is a short distance indeed from entities and processes to formulas and rules when we are considering all four as instances of types of empirical discoveries in science.

Formulas are shorthand statements of factual relations which can be readily applied by substituting particular data. Usually expressed as mathematical equations, when employed as guides to procedure they call for reactions by substituting concrete values for variables. In chemistry, for instance, formulas for compounds consist in symbols representing the ingredient elements. The formula for magnesium chloride is MgC_2, one part of magnesium to two parts of chlorine.

A rule is a formula for a specific operation on specific material objects, a principle of practice or procedure. A well known rule in chemistry is Gibbs' "phase rule," according to which for any heterogeneous system in equilibrium the sum of the number of phases (i.e. separate parts of a heterogeneous system) plus the number of degrees of freedom (i.e. the least number of independent variables) is equal to the number of components (i.e. the least number of different substances) plus two.

It often happens that the terms, formulas and rules, are used interchangeably. In physics, that the speed of light cannot be exceeded by the speed of any material body may be considered a formula. On the other hand, any standardized chemical reaction may be stated as a rule. Generally speaking, formulas are more technical than rules; formulas describe mathematical relationships, while rules direct laboratory operations. One speaks of rules of inference and of rules of induction; but these are logical and mathematical, not empirical. The term, in experimental science is more often than not employed in connection with short-range operational directives, laboratory "rules-of-thumb." They are tiny invariants, the results of trial-and-error which may have been verified empirically but which have no support in more comprehensive abstract theory.

7. *Procedural principles*

In addition to the empirical systems and areas, the laws, entities and processes, which make up the bulk of the findings of empirical science, there are also those findings which consist in procedural principles. There is only one scientific method, but it is applied with appropriate modifications which adapt it to the needs of particular sciences. Thus the use of control groups is a common method in biological investigation but certainly would not be possible in astronomy. Procedural principles are maxims of investigation appropriate to a given area; usually they are somewhat less wide than the science in which they apply. A good example is the uncertainty principle in quantum mechanics or the correspondence principle. The principle of uncertainty, that either the position or the velocity of a particle can be determined but not both, began as a procedural principle. It has since been claimed in quantum mechanics that this principle represents the constitutive as well as the regulative character of certain subatomic phenomena.

There is a thin line indeed between a procedural principle and a predictive law in an advanced science. The correspondence principle, which states that the statistical quantum laws must on an average lead to the classical equations as a limit, has an element of legalistic prediction about it. Still, it is possible to distinguish between the correspondence principle, say, and the mass-energy conversion formula. Any law in a science may be employed as a procedural principle. The law (so-called) of constant proportions in chemistry, according to which all samples of a substance contain elements in the same proportions, may be interpreted as a procedural principle.

There is a sense in which all of the mathematical systems which have been applied in the sciences may be regarded as sets of procedural principles, They direct what has to be done in order to obtain particular results. The branch of mathematics called statistical probability is a case in point. Statistical probability, one might say, consists in a set of procedural principles for treating masses of experimental data. For the most part, however, what is meant by procedural principles are the principles whereby those data may be amassed.

We have said that procedural principles are usually those which are peculiar to the methodology of a given science, and more often to some specific area in a science. There are generalized scientific procedural principles as well. Such a one is the necessity to modify laws in their application by means of extenuating circumstances. Gravity calculations have to be modified in terms of the density of the atmosphere, for instance. The structure of the scientific method of investigation might be described as an ordered assemblage of procedural principles. That these are often encountered in actual experiments or calculations makes them a type of empirical discovery. The self-corrective nature of the scientific method applies to the fashion in which that method continually effects its own improvement.

8. *The limits of empirical discovery*

The aim of empirical discovery is of course the totality of knowledge concerning the world according to science. The nearer empirical systems approach abstract mathematical systems and the nearer the laws approach pure mathematical equations, the clearer it becomes that there are limits to scientific discovery set by the limiting conditions of the empirical world. The pre-scientific philosophies were extrapolated at will in terms of universals without regard to observation or experiment, prediction or control; but science requires its generalities to have some empirical support. Generalities they are, however, none the less, and it must be remembered that universals are after all only extreme cases of generalities; if understood in the proper context, they are the ideal limiting cases. Scientific investigation comes to an end with the boundary conditions. These do not exist as material objects, but subsist, and they operate from a logical status; they continue to offer the twin advantages of knowledge of nature and control over nature.

The totality of world conditions cannot be known. For to know them would mean to construct the widest system. What could be used for scaffolding in the course of constructing the consistency and completeness of such a system? The task belongs to a meta-system, and this would mean, then, that the system itself was not the widest system. We are presented with the prospect of an infinite regress of systems. There is after all the well known theorem of Gödel, according to which the consistency of any system (without which it is not a system) cannot be proved within the system itself. But to prove it from without requires a meta-system again. If it is true that the greatest discovery of the scientific age has been the method of discovery, then it is true also that the most sobering discovery has been the limits of discovery: there is no such thing as the widest system and hence no knowledge is possible of the totality of world conditions.

The objects available to scientific investigation are always open and the systems by means of which the investigation is conducted are always closed. Scientists can hope to learn as much about the world as their techniques allow. It remains only to admit, then, that the world contains more than is represented in any system of ideas, no matter how complete and consistent such a system may be. Scientific formulations are constantly being revised in view of newly acquired knowledge, and so more and more of the world is represented; but there is always still more of the world. What is needed at this point, then, is a principle of limitation. The principle of fallibility is also a principle of limitation, and relies upon the truth of the proposition that the world is qualitatively richer and structurally more complex than any scientific description. And this principle itself must be included in every comprehensive formulation.

The typical findings of science are not therefore completely inclusive empirical systems but laws. Laws are what science seeks. The discovery of empirical areas means for science new opportunities to win the knowledge of laws. Entities and processes are but the constants and variables which are found to be combined into laws when their relations are made known. Science proceeds on a much more humble method than the all-presumptive aim of total knowledge. It is content to discover a law.

Within empirical areas, then, and dealing with actual entities and processes, the investigator is chiefly concerned with finding laws. Laws experimentally verifiable rather than empirical systems are the archetype of scientific discoveries. They are the typical invariants in the search among actual material elements, the permanent abstract relationships which, while themselves independent of everything actual, everything in actuality exemplifies. For while everything actual changes, what changes may change at an unchanging rate. In this sense all laws serve as models, and events conform to their exemplification.

Laws combine into systems, and by seeing how they do so we can distinguish at least two varieties. There are the laws fundamental *to* a system and the laws formulated *within* a system. Laws hold as laws when they are considered in their relation to the relevant data. This is a question of correspondence; but there is also the question of coherence. Laws are also the components of abstract empirical systems. There can be no better example of laws fundamental to a system than the three laws of energy in thermodynamics. An equally good example of a law formulated within a system is Coulomb's Law for the measurement of the attraction of electrical forces.

Depending upon the frame of reference, laws can be regarded as quasi-independent or as elements within a system. Weizsäcker (1952), for instance, has shown how Coulomb's Law can be considered a law of field physics by reformulating it: the electrical field strength is inversely proportional to the square of the distance from a point of charge in the neighborhood. There are no absolutely self-contained scientific formulations; which is a negative way of asserting that underlying the scientific findings there is the silent postulate of the unity of nature. As Bohm has shown (1951), at the quantum level there

are no absolutely intrinsic properties of material objects. Particles share their properties with the systems with which they interact. No wonder, then, that in the last analysis laws are not valid isolates in any ultimate sense. As we noted at the beginning of this section, there are definite limits in the direction of knowledge of the totality of world conditions. The laws possess such validity as there is. And we may for this purpose consider empirical systems as made up of them, and their discovery as the business of the scientific method.

One final word. This concerns what for want of a better term may be called empirical detachment, the separation from science of the discoveries made by science with which they have no essential or necessary connection. It often happens that science is confused with many of those scientific findings which otherwise would not be known. But the findings themselves are not inherently "scientific," they are facts of nature. Thus that there is cosmic radiation as a discovery of astrophysics, and that there are helical genetic chains was a discovery of molecular biology. But cosmic radiation and helical genetic chains are ingredients of the natural world and existed no doubt millenia before man learned about them. What astrophysics and molecular biology discover are not inherently physical or biological in the scientific sense but part of the body of reliable knowledge about the world which man has come to know and in time may also use for his own ends. Science is only an inquiry, and does not permanently retain the results of that inquiry, even though it is often only by means of the inquiry that such results are ever known. Science increases man's knowledge of the world and then gracefully steps out of the way in order that such knowledge may aid his understanding or his practice without further intermediation.

REFERENCES

Bavink, B. *The Natural Sciences.* (New York 1932, Century).

Benjamin, A. C. *The Logical Structure of Science.* (London 1936, Kegan Paul).

— *An Introduction to the Philosophy of Science.* (New York 1937, Macmillan).

Bernard, Claude. *An Introduction to the Study of Experimental Medicine.* H. C. Green trans. (New York 1949, Schuman).

Bethe, H. A. *Elementary Nuclear Theory.* (New York 1947, Wiley).

Black, M. *The Nature of Mathematics.* (New York 1934, Harcourt Brace).

Bohm, David. *Quantum Theory.* (Englewood Cliffs, N. J. 1956, Prentice-Hall).

— *Causality and Chance in Modern Physics.* (Princeton 1957, Van Nostrand).

Born, M. *Natural Philosophy of Cause and Chance.* (Oxford 1949, Clarendon Press).

Carnap, Rudolf. *Logical Foundations of Probability.* (Chicago 1950, University Press).

— *Philosophical Foundations of Physics.* (New York 1966, Basic Books).

Churchman, C. W. and Ackoff, R. L. *Methods of Inquiry.* (St. Louis 1950, Educational Publishers, Inc.).

Chwistek, Leon. *The Limits of Science.* (London 1948, Kegan Paul).

Clark, N. (ed.) *A Physics Anthology.* (London 1960, Chapman and Hall).

Clifford, W. K. *The Common Sense of the Exact Sciences.* (New York 1946, Knopf).

Cohen, M. R. *Reason and Nature.* (New York 1931, Harcourt Brace).

— and Nagel, E. *An Introduction to Logic and Scientific Method.* (New York 1934, Harcourt Brace).

Condon, E. U. and Odishaw, H. *Handbook of Physics.* (New York 1958, McGraw-Hill).

Copi, Irving M. *Introduction to Logic*. Third Edition. (New York 1968, Macmillan).

Cournot, Antoine Augustin. *An Essay on the Foundations of Our Knowledge*. M. H. Moore trans. (New York 1956, Liberal Arts Press).

Cundy, H. M. and Rollett, A. P. *Mathematical Models*. (Oxford 1952, Clarendon Press).

Darwin, C. *Autobiography*. (London 1958, Collins).

DeBroglie, Louis. *Physics and Microphysics*. (London 1955, Hutchinson).

— *The Revolution in Physics*. R. W. Niemeyer trans. (New York 1953, Noonday Press).

Duhem, P. *The Aim and Structure of Physical Theory*. (Princeton 1954, University Press).

Eaton, R. M. *General Logic*. (New York 1931, Scribner).

Einstein, Albert. *Relativity: The Special and General Theory*. (New York 1920, Peter Smith).

— et al. *The Principle of Relativity*. (London 1923, Methuen).

— *The World As I See It*. (New York 1932, Covict Friede).

— *The Meaning of Relativity*. (Princeton 1945, University Press.)

— *The Meaning of Relativity*. Third Edition. (Princeton 1950, University Press).

— *Out of my Later Years*. (New York 1950, Philosophical Library).

— and Infield, L. *The Evolution of Physics*. (New York 1942, Simon and Schuster).

Ellis, Brian. *Basic Concepts of Measurement*. (Cambridge 1966, University Press).

Faraday, M. *Faraday's Diary*. 7 Volumes. (London 1932, Bell and Sons).

— *Experimental Researches in Chemistry and Physics*. (London, 1846).

— *Philosophical Transactions*. (London).

Feynman, Richard. *The Character of Physical Law* (Cambridge, Mass., 1956, M.I.T. Press).

Fisher, R. A. *Statistical Methods for Research Workers* (London 1938, Oliver and Boyd).

— *The Design of Experiments*. (London 1949, Oliver and Boyd).

— *Statistical Methods and Scientific Inference*. (London 1956, Oliver and Boyd).

Freedman, Paul. *The Principles of Scientific Research*. (London 1949, MacDonald).

Friend, J. W. and Feibleman, J. K. *What Science really Means*. (London 1937, Allen and Unwin).

Galilei, G. *Dialogues Concerning Two New Sciences*. (Evanston 1946, Northwestern University).

Hanson, N. R. *Patterns of Discovery.* (Cambridge 1958, University Press).
Harrod, R. *Foundations of Inductive Logic.* (New York 1956, Harcourt Brace).
Heisenberg, Werner. *Philosophical Problems of Nuclear Science.* F. C. Hayes trans. (New York 1952, Pantheon).
— *Physics and Philosophy.* (New York 1958, Harper).
Hempel, Carl G. *Aspects of Scientific Explanation.* (New York 1965, Free Press).
Henkin, L. et al. *The Axiomatic Method.* (Amsterdam 1959, North-Holland Publishing Company).
Herschel, J. F. W. *A Preliminary Discourse on the Study of Natural Philosophy.* (Philadelphia, 1831, Carey and Lea).
Hilbert, D. and Ackermann, W. *Principles of Mathematical Logic.* (New York 1950, Chelsea).
Hutten, Ernest H. *The Language of Modern Physics.* (London 1956, Allen and Unwin).
Jammer, Max. *The Conceptual Development of Quantum Mechanics.* (New York 1966, McGraw-Hill).
Jeffreys, Harold and Jeffreys, Bertha Swirles. *Methods of Mathematical Physics.* (Cambridge 1946, University Press).
Jevons, W. S. *The Principles of Science.* (London 1879, Macmillan).
— *The Theory of Political Economy.* (London 1924, Macmillan).
Kattsoff, L. O. *A Philosophy of Mathematics.* (Ames, Iowa 1948, Iowa State College Press).
Kemeny, J. G. *A Philosopher Looks at Science.* (Princeton 1959, Van Nostrand).
Keynes, J. M. *A Treatise on Probability.* (London 1948, Macmillan).
Kolmogorov, A. N. *Foundations of the Theory of Probability.* (New York 1956, Chelsea).
Körner, S. (ed.) *Observation and Interpretation.* (New York 1957, Academic Press).
Lenzen, V. F. *Causality in Natural Science.* (Springfield, Ill. 1954, Thomas).
Levy, H. *The Universe of Science.* (New York 1933, Century).
Lindsay, R. B. and Margenau, H. *Foundations of Physics.* (New York 1957, Dover).
Lorentz, H. A. and Einstein, A. et al., *The Principles of Relativity.* (New York, no date, Dover).
Mach, E. *The Analysis of Sensations.* (Chicago 1914, Open Court).
Madden, E. H. (ed.) *The Philosophical Writings of Chauncey Wright.* (New York 1958, Liberal Arts Press).

Magoun, H. W. *The Waking Brain.* (Springfield, Ill. 1960, Charles C. Thomas).

Mandl, F. *Quantum Mechanics.* (London 1957, Butterworth).

Margenau, Henry and Murphy, George Moseley. *The Mathematics of Physics and Chemistry.* (Princeton 1956, Van Nostrand).

Martin, R. M. *Truth and Denotation.* (London 1958, Routledge and Kegan Paul).

Meyerson, Emile. *Identity and Reality.* Kate Loewenberg trans. (London 1930, Allen and Unwin).

Mill, J. S. *A System of Logic.* (London 1936, Longmans Green).

Morgenbesser, Sidney. (ed.) *Philosophy of Science Today.* (New York 1967, Basic Books).

Nagel, Ernest. *The Structure of Science.* (New York 1961, Harcourt Brace and World).

Newton's Principia. Trans. by Motte and revised by Cajori. (Berkeley 1946, University of California Press).

Oparin, A. I. *The Origin of Life on the Earth.* (New York 1957, Academic Press).

Pap, Arthur. *An Introduction to the Philosophy of Science.* (Glencoe, Ill. 1962, Free Press).

Peirce, C. S. *Chance, Love and Logic.* (M. R. Cohen, Ed.). (London 1932, Kegan Paul).

— *Collected Papers.* (Cambridge 1931-35, Harvard University Press).

— *Collected Papers.* 8 Volumes. (Cambridge 1931-58, Harvard University Press).

— *Collected Papers,* Volume IV. (C. Hartshorne and P. Weiss, eds.). (Cambridge 1933, Harvard University Press).

— *Letters to Lady Welby.* (I. C. Leib, ed.). (New Haven 1953, Whitlock).

Planck, M. *The Universe in the Light of Modern Physics.* (New York 1931, Norton).

— *Where is Science Going?* (London 1931, Allen and Unwin).

— *The Philosophy of Physics.* (London 1936, Allen and Unwin).

Poincaré, H. *The Foundations of Science.* (New York 1929, Science Press).

Polanyi, Michael. *Personal Knowledge.* (Chicago 1958, University Press).

Popper, Karl R. *The Logic of Scientific Discovery.* (London 1959, Hutchinson).

— *Conjectures and Refutations.* (London 1963, Kegan Paul).

Quine, W. V. O. *From a Logical Point of View.* (Cambridge 1953, Harvard University Press).

Russell, Bertrand. *The Principles of Mathematics.* (London 1937, Allen and Unwin).

— *Human Knowledge.* (New York 1948, Simon and Schuster).
Schlick, M. *Gesammelte Aufsätze.* (Wien 1938, Gerold and Company).
Schlipp, Paul A. *Albert Einstein: Philosopher-Scientist.* (Evanston, Ill. 1949, Library of Living Philosophers).
Searles, H. L. *Logic and Scientific Methods.* (New York 1948, Ronald Press).
Shamos, M. H. *Great Experiments in Physics.* (New York 1959, Holt).
Singer, Charles. et al. *A History of Technology.* 5 Volumes. (London 1954, Oxford).
Smithsonian Institute Report for 1951. (Washington 1952, United States Government Printing Office).
Stebbing, L. S. *A Modern Introduction to Logic.* (New York 1930, Crowell).
Strong, John. et al. *Procedures in Experimental Physics.* (New York 1938, Prentice-Hall).
Tarski, A. *Introduction to Logic.* (New York 1946, Oxford University Press).
Tinbergen, N. *The Herring Gull's World.* (London 1953, Collins).
Trotter, W. *The Collected Papers of Wilfred Trotter.* (London 1946, Oxford University Press).
Van 't Hoff, J. H. *Imagination in Science.* Molecular Biology 1. (New York 1967, Springer-Verlag).
von Mises, R. *Probability, Statistics and Truth.* (New York 1939, Macmillan).
von Weizsäcker, C. F. and Juilfs, J. *Contemporary Physics.* (London 1957, Hutchinson).
von Wright, G. H. *A Treatise on Induction and Probability.* (London 1951, Routledge and Kegan Paul).
Waddington, C. H. *The Scientific Attitude.* (Middlesex 1948, Pelican).
Wartofsky, Mark W. *Conceptual Foundations of Scientific Thought.* (New York 1968, Macmillan).
Watson, W. H. *On Understanding Physics.* (Cambridge 1938, University Press).
Werkmeister, W. H. *A Philosophy of Science.* (New York 1940, Harper and Bros.).
— *The Basis and Structure of Knowledge.* (New York 1948, Harper and Bros.).
Weyl, Hermann. *The Theory of Groups and Quantum Mechanics.* (New York, no date, Dover).
Whitehead, A. N. *The Concept of Nature.* Chapter IV. (Cambridge 1936, University Press).
— *The Principles of Natural Knowledge.* Part III. (Cambridge 1955, University Press).

— and Russell, B. *Principia Mathematics*. 3 Volumes. (Cambridge 1925, University Press).

Wilson, E. Bright. *An Introduction to Scientific Research*. (New York 1952, McGraw-Hill).

Wisdom, John O. *Foundations of Inference in Natural Science*. (London (1952, Methuen).

Wittgenstein, L. *Tractatus Logico-Philosophicus*. (London 1933, Kegan Paul).

Worthing, Archie G. and Geffner, Joseph. *Treatment of Experimental Data*. (New York 1944, Wiley).

Wyckoff, Ralph W. G. *The World of the Electron Microscope*. (New Haven 1958, Yale University Press).

Yule, G. U. and Kendall, M. G. *An Introduction to the Theory of Statistics*. (Philadelphia 1940, Lippincott).

INDEX